Lecture Notes in Computer Science

Commenced Publication in 1973
Founding and Former Series Editors:
Gerhard Goos, Juris Hartmanis, and Jan van Leeuwen

Steve Uhlig   Konstantina Papagiannaki
Olivier Bonaventure (Eds.)

Passive and Active Network Measurement

8th International Conference, PAM 2007
Louvain-la-Neuve, Belgium, April 5-6, 2007
Proceedings

 Springer

Volume Editors

Steve Uhlig
Delft University of Technology
4 Mekelweg, 2628 CD Delft, The Netherlands
E-mail: s.p.w.g.uhlig@ewi.tudelft.nl

Konstantina Papagiannaki
Intel Research Pittsburgh
4720 Forbes Avenue, Suite 410, Pittsburgh, PA 15213, USA
E-mail: dina.papagiannaki@intel.com

Olivier Bonaventure
Université catholique de Louvain
2, Place Sainte Barbe, 1348 Louvain-la-Neuve, Belgium
E-mail: Olivier.Bonaventure@uclouvain.be

Library of Congress Control Number: 2007923176

CR Subject Classification (1998): C.2, C.4, H.4, K.6.5

LNCS Sublibrary: SL 5 – Computer Communication Networks and Telecommunications

ISSN   0302-9743
ISBN-10   3-540-71616-5 Springer Berlin Heidelberg New York
ISBN-13   978-3-540-71616-7 Springer Berlin Heidelberg New York

Springer is a part of Springer Science+Business Media

springer.com

© Springer-Verlag Berlin Heidelberg 2007

Typesetting: Camera-ready by author, data conversion by Scientific Publishing Services, Chennai, India
Printed on acid-free paper   SPIN: 12041985   06/3142   5 4 3 2 1 0

Preface

The 2007 edition of the Passive and Active Measurement Conference is the eighth of a series of successful events. Since 2000, the Passive and Active Measurement (PAM) conference has provided a forum for presenting and discussing innovative and early work in the area of Internet measurement. This event focuses on the research and practical applications of network measurement and analysis techniques. The conference's goal is to provide a forum for current work in its early stages. This year's conference was held in Louvain-la-Neuve, the youngest city in Belgium, built in the 1970s to host the new campus of the Université catholique de Louvain, one of the oldest universities in Europe.

The call for papers attracted 80 submissions. Each paper was carefully reviewed by three members of the Technical Program Committee. The reviewing process led to the acceptance of 21 full papers and 12 short papers. The papers were arranged into seven regular and one poster session covering the following areas: interdomain routing, P2P, wireless 802.11, wireless 3G/CDMA/Bluetooth, infrastructure and services, traffic, and measurement principles. Similarly to previous years, we saw from the technical sessions the emerging importance of wireless and applications in the topics covered by the conference. This is a sign that PAM authors indeed work on timely issues related to Internet measurements. The technical program of the conference was complemented by a half-day PhD student session with invited presentations and a panel.

This year, the PC Chair, Konstantina Papagiannaki, took the initiative of trying an experiment on review quality. After this year's PAM program had been finalized, we asked authors to rank the quality of each review they received. Participation in the scheme was optional both for authors and reviewers and did not affect the decision for any paper. The intent of this experiment was to help us understand how authors value reviews, and to provide TPC members with feedback on the reviews they have written. The findings produced by the analysis of the feedback received will be published in the July issue of the *ACM Computer Communications Review.*

The organization of such a conference would not have been possible without the help of many persons. In particular, we would like to thank Stéphanie Landrain for her hard work in organizing the social events, handling all registrations, and taking care of all the important details. We would also like to thank Benoit Donnet, who managed the Web site during the last months and helped to solve many problems.

We are very grateful to Endace, Intel and Cisco Systems whose sponsoring allowed us to keep low registration costs and also to offer several travel grants to PhD students.

April 2007

Steve Uhlig
Konstantina Papagiannaki
Olivier Bonaventure

Organization

Organization Committee

General Chair	Steve Uhlig (Delft University of Technology, The Netherlands)
Program Chair	Konstantina Papagiannaki (Intel Research Pittsburgh, USA)
Local Arrangements Chair	Olivier Bonaventure (Université catholique de Louvain, Belgium)
Finance Chair	Stéphanie Landrain (Université catholique de Louvain, Belgium)

Program Committee

Mark Allman	ICSI, USA
Suman Banerjee	University of Wisconsin, Madison, USA
Ethan Blanton	Purdue University, USA
Nevil Brownlee	University of Auckland, New Zealand
Mark Claypool	WPI, USA
Christophe Diot	Thomson Research, Paris, France
Christos Gkantsidis	Microsoft Research, Cambridge, UK
Gianluca Iannaccone	Intel Research Berkeley, USA
Balachander Krishnamurthy	AT&T Research, USA
Simon Leinen	SWITCH, Switzerland
Bruce Maggs	CMU/Akamai Technologies, USA
Ratul Mahajan	Microsoft Research, USA
Alberto Medina	BBN Technologies, USA
Konstantina Papagiannaki	Intel Research Pittsburgh, USA
Lili Qiu	University of Texas at Austin, USA
Coleen Shannon	CAIDA, USA
Peter Steenkiste	CMU, USA
Steve Uhlig	Delft University of Technology, The Netherlands
Jia Wang	AT&T Research, USA
David Wetherall	University of Washington, USA
Tilman Wolf	University of Massachusetts, Amherst, USA

Steering Committee

Mark Allman	ICSI, USA
Chadi Barakat	INRIA, France
Nevil Brownlee	University of Auckland, New Zealand

Constantinos Dovrolis   Georgia Tech, USA
Ian Graham   Endace, New Zealand
Konstantina Papagiannaki   Intel Research Pittsburgh, USA
Matthew Roughan   University of Adelaide, Australia
Steve Uhlig   Delft University of Technology,
                                The Netherlands

Sponsoring Institutions

Endace
Cisco Systems
Intel Corp.

Table of Contents

Measures of Self-similarity of BGP Updates and Implications for Securing BGP

Geoff Huston

Centre for Advanced Internet Architectures,
Swinburne University of Technology,
Melbourne, Australia
gih@swin.edu.au

Abstract. Techniques for authenticating BGP protocol objects entail the inspection of additional information in the form of authentication credentials that can be used to validate the contents of the BGP update message. The additional task of validation of these credentials when processing BGP messages will entail significant additional processing overheads. If the BGP validation process is prepared to assume that a validation outcome has a reasonable lifetime before requiring re-validation, then a local cache of BGP validation outcomes may provide significant leverage in reducing the additional processing overhead. The question then is whether we can quantify the extent to which caching of BGP updates and the associated validation outcome can reduce the validation processing load. The approach used to address this question is to analyze a set of BGP update message logs collected from a regional transit routing location within the public IPv4 Internet. This paper describes the outcomes of this study into the self-similarity of BGP updates and relates these self-similarity metrics to the size and retention time characteristics of an effective BGP update cache. This data is then related to the message validation activity, and the extent to which caching can reduce this validation processing activity is derived.

Keywords: BGP, Secure BGP, Validation Caching.

1 Introduction

The scaling properties of BGP [1] have represented a long-term concern for the viability of BGP in the role of supporting the inter-domain routing system of the public Internet. These concerns were first raised in the early 1990's [2] and continue to the present time. The elements of this concerns are twofold: firstly, the 'size' of the routing space, as expressed by the number of discrete entries to be found in a default-free BGP routing table, and, secondly, the rate of BGP Updates messages to be processed, which relates to the 'processing load' of the routing system. A study of the characteristics of BGP update messages over the entire year of 2005 indicated that the number of discrete entries in the BGP routing table grew by 18% to a total at the end of the year of some 175,400 entries, the number of per-prefix updates per eBGP peer session grew by 49% over the same period, and the number of BGP prefix withdrawals grew by 112% [3]. The salient observation here is that the number of BGP updates, or the

S. Uhlig, K. Papagiannaki, and O. Bonaventure (Eds.): PAM 2007, LNCS 4427, pp. 1–10, 2007.

'processing load' associated with BGP in the Internet is growing at a higher rate than the number of BGP entries, or the 'size' of the BGP routing table. The implication here is that 'processor load' appears to represent the more critical scaling factor for BGP in the context of the growth trends within the public Internet.

Adding additional attributes to the BGP protocol that are intended to improve the security of BGP also have the potential to impose higher processing loads per BGP update message. The significant factor in incremental processing loads is that associated with the task of validating the contents of the BGP update message. The basic security questions with respect to securing the BGP update payload include validation of the address prefix, validation of the origin AS, validation that the origin AS has the authority to originate an advertisement of this prefix, and validating the AS Path attribute of the update as being an accurate representation of the advertised path to reach this address prefix [4].

Irrespective of the manner by which the authentication credentials associated with a BGP update are propagated across the routing realm, the security consideration is that the BGP receiver should validate these credential as a precursor to accepting the BGP update as authentic. Approaches to validation of such information commonly rely on public key cryptography, where the original 'author' of the information can attach a signature to the information using their private key, and any recipient of the information can validate the authenticity of the information by validating the signature block through using the matching public key. Such a validation process implies some form of additional processing load.

One approach to minimize the additional processing overhead associated with validation of BGP updates is to use validation caching. In this approach the validation outcome of the BGP update is cached for a period of time, and if the update is repeated within this period, then the previous validation outcome is used without further validation checking.

The questions that this approach raise include: How effective could validation caching be in this context? How big a cache is appropriate in terms of size and hit-rate trade-offs? What time-period of validation caching would be appropriate in terms of a balance between cache hit rates and validation accuracy?

The remainder of this paper is structured as follows. Section 2 provides a description of the measurement and analysis methodology used in this study, and the data set used by this study. Section 3 describes the outcomes of this study, relating self-similarity results to per-update validation processing loads. Section 4 concludes the paper.

2 Methodology

The Zebra implementation of BGP [5] was configured as a BGP update message collector, and this collector was configured with a single eBGP session to AS4637. The motivation for this simple BGP configuration was to isolate the update message sequence that correspond to a single eBGP peering session, and for the update message sequence to reflect the changes that occur in the Local Routing Information Base (LOC-RIB) of the peer AS.

The configuration was set up to collect the complete set of BGP update messages, place a timestamp on each update and save then on a rolling 24 hour basis. Also, a snapshot was taken of the BGP routing table at the end of each 24 hour cycle. This collection process commenced on the 3rd March 2006, and the data set, continuously updated on a daily basis, has been published as a data resource.

For this study, the data corresponding to a two week period from midnight 10th September 2006 to 23:59 23rd September 2006[1,2] was used. The update log data from the BGP collection unit was translated to a time-sequence of per-prefix update transactions, reflecting the sequence and time of changes to the LOC-RIB of the BGP speaker in the peer AS.

This sequence of BGP transactions was used as input into a multi-dimension cache simulator. This simulator simultaneously simulates a set of fix-length caches ranging in size from 1 through to the number of unique updates, and reports on the cache hit rate for each possible size of the cache.

3 Measurement Results

In the 14 day analysis period there were a total of 656,339 BGP Update messages. The profile of the BGP state across this 14 day period is shown in Table 1.

Table 1. BGP Profile for the period 10-September 2006 00:00 to 23-September-2006 23:59

Metric	Value
Number of BGP Update Messages	656,339
Prefix Updates	1,632,900
Prefix Withdrawals	223,616
Average BGP Update Messages per second	0.54
Average Prefix Updates per second	1.53
Peak Prefix Update Rate per second	4999
Prefix Count	202,769
Updated Prefix Count	111,769
Stable Prefix Count	91,000
Origin AS Count	23,233
Updated Origin AS Count	15,501
Stable Origin AS Count	7,732
Unique AS Path Count	87,238
Updated Path Count	75,529
Stable AS Paths	11,709

The distribution of updates is not uniform across the set of prefixes, nor is it uniform across the set of origin ASs. The most unstable 1% of prefixes generated 18% of the total number of BGP prefix updates, and the top 50 prefixes (0.025% of the prefixes in the BGP routing table) generated 3.5% of all prefix updates. A similar skewed distribution is evident for autonomous systems, where the busiest 1% of the Origin ASs (232 ASs) were affected by 38% of the total updates, and the top 50 Origin ASs (0.2% of the ASs) were involved in 24% of all the BGP prefix updates.

---

[1] The times quoted here refer to the local time at UTC+10 hours.
[2] The BGP data set is published at http://www.potaroo.net/papers/phd/pam-2007/data

In considering the potential to cache validation outcomes of BGP updates, it is noted that the validation of a BGP Update has a time component, in that the validation outcome does not remain useable indefinitely. The potential effectiveness of validation caching depends on both the time characteristics of self-similar BGP updates (the elapsed time between identically keyed updates), as well as their space characteristics (the number of different updates between identically keyed updates).

Table 2. BGP Self-Similarity Total Counts per day

Day	Prefix Updates	Duplicates: Prefix	Duplicates: Prefix + Origin AS	Duplicates Prefix + AS Path	Duplicates Prefix + Norm-Path
1	72,934	60,105 (82%)	54,924 (75%)	34,822 (48%)	35,312 (48%)
2	79,361	71,714 (90%)	67,942 (86%)	49,290 (62%)	50,974 (64%)
3	104,764	93,708 (89%)	87,835 (84%)	65,510 (63%)	66,789 (64%)
4	107,576	94,127 (87%)	87,275 (81%)	64,335 (60%)	66,487 (62%)
5	139,483	110,994 (80%)	99,171 (71%)	68,096 (49%)	69,886 (50%)
6	100,444	92,944 (92%)	88,765 (88%)	70,759 (70%)	72,108 (72%)
7	75,519	71,935 (95%)	69,383 (92%)	56,743 (75%)	58,212 (77%)
8	64,010	60,642 (95%)	57,767 (90%)	49,151 (77%)	49,807 (78%)
9	94,944	89,777 (95%)	86,517 (91%)	71,118 (75%)	72,087 (76%)
10	81,576	78,245 (96%)	75,529 (93%)	63,607 (78%)	64,696 (79%)
11	95,062	91,144 (96%)	87,486 (92%)	72,678 (76%)	74,226 (78%)
12	108,987	103,463 (95%)	99,662 (91%)	80,720 (74%)	82,290 (76%)
13	91,732	87,998 (96%)	85,030 (93%)	72,660 (79%)	74,116 (81%)
14	78,407	76,174 (97%)	74,035 (94%)	64,994 (83%)	65,509 (84%)

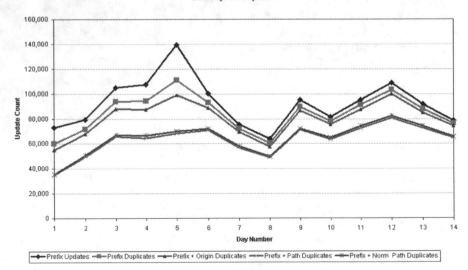

Fig. 1. Daily Duplicate Update Profile

In this study four types of self-similarity are considered, namely the recurrence of updates that specify the same address prefix, secondly, updates that share the same address prefix and origin AS, thirdly, the same address prefix and AS Path, and finally updates that share the same address prefix and the same 'normalized' AS Path (in this context 'normalization' of an AS Path implies the removal of duplicate AS numbers from the AS Path that are the result of path pre-pending).

The occurrence of self-similar updates across the 14 day period is indicated in Table 2 ("Norm-Path" in this table refers to equivalence using the 'Normalized' AS Path). The daily totals are plotted in Figure 1.

3.1 Time Distribution of BGP Updates

The first question concerns the time spread of self-similar BGP updates and concerns the distribution of time intervals between pairs of similar BGP updates, using the four types of prefix update similarity noted in the previous section. While the time resolution of the collected update log data is in units of milliseconds, the eBGP session used for data collection uses a 30 second value for the min-route-advertisement timer, so that all updates are passed to the collection unit in 30 second intervals. Accordingly, in this examination of the time spread of self-similar BGP updates, the time intervals are aggregated into 30 second increments.

A cumulative histogram of the proportion of recurring update messages by varying recurrence intervals of these four types of update self-similarity are shown in Figure 2. The actual number of duplicate updates are shown in Figure 3. Some 49% of all prefix-recurring updates occur within 90 seconds of the previous update. Some of these high frequency updates could be due to BGP convergence behaviour following a prefix withdrawal, where BGP will explore a number of alternative routes before the withdrawal has been propagated across the entire network. When looking at

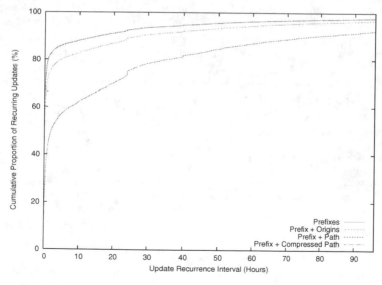

Fig. 2. Cumulative Proportion of Recurring Updates for Address Prefixes

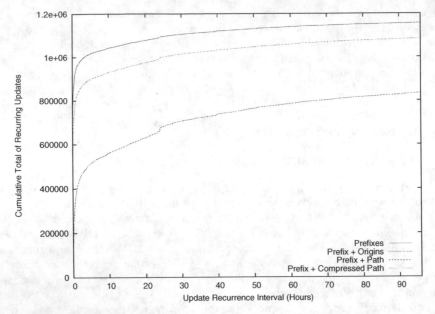

Fig. 3. Cumulative Volume of Recurring Updates for Address Prefixes

recurrence of prefix + path only 8% of all similar updates occur within the same 90 second interval. As BGP withdrawal-triggered convergence should not explore the same route twice, the relative difference the two figures indicate that up to 40% of the recurring BGP updates that refer to the same prefix may be attributable to short term updates generated during the process of BGP convergence.

Some 80% of recurring prefix updates occur within 1 hour of the preceding prefix update. The same 80% threshold occurs within 35 hours when considering the recurrence of BGP updates that contain the same prefix and AS Path. There is little difference between Path and Compressed Path similarity measures in either short (1 hour) or longer (36 hour) timeframes.

If a benchmark value for potential cache efficiency is set to 80%, and the validation cache process is intended to cache the outcome of both prefix origination and AS Path validity, then a validation retention time period of 36 hours would be a minimum value to achieve this performance metric.

3.2 Validation Cache Simulations

The next metric concerns the spatial dispersion of similar BGP updates, where the number of intervening unique updates between two matching updates is considered. This dispersion relates to the size of a cache that would be able to generate a cache hit for the matching updates when using a simple Least Recently Used (LRU) cache management regime.

Figure 4 shows the cache hit rate per day for each of the 14 days in the study period, using a simple LRU cache scheme. This data indicates that a cache of 200 prefixes would provide an average hit rate of between 50 to 70%. Improvement in caching efficiency per incremental unit of cache size appears to drop once the cache

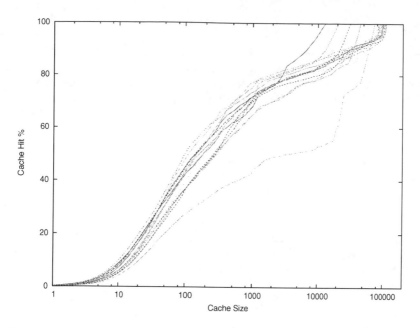

Fig. 4. Cache Hit rates per Day for a range of Cache Sizes. The cache algorithm is LRU, using a lookup key of the prefix.

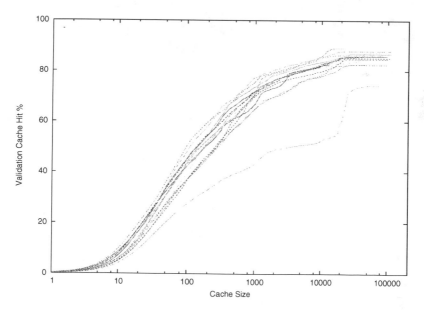

Fig. 5. Validation Cache Hit rates per Day for a range of Cache Sizes. The cache algorithm is LRU, using a lookup key of the prefix, with a validation re-use period of 36 hours.

size exceeds 1000. The outlier 24 hour period sequence correlates to a BGP reset of a transit peering session which, in turn, generated a large number of one-off prefix updates, which caused the lower than average cache hit rate for this 24 hour period.

If authentication data were applied to prefixes, and a BGP receiver validated this prefix data, then it is possible to look at the potential improvements that a prefix validation cache could provide. This is shown in Figure 5. It is noted that this data set uses a validation lifetime of 36 hours, such that a cached entry will be considered stale and re-validation required once the entry is more than 36 hours old. Within each 24 hour period some 15% to 20% of announced prefixes are announced more than 36 hours after their previous announcement, while some 80% of announced prefixes have been previously announced less than 36 hours previously. The overall majority of announced prefixes in BGP are short lived announcements that are refined by subsequent updates within the ensuing 36 hours. This data indicates that a per-eBGP peer validation cache of 1000 prefixes, managed on an LRU basis with a 36 hour validation period would provide a reduction in validation processing of BGP updates by between 60% to 80% on a daily basis. Increasing this cache to 10,000 entries could improve this validation processing load reduction to between 80% to 90%.

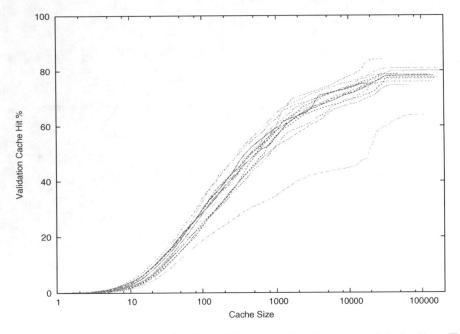

Fig. 6. Prefix + Origin AS Validation Cache Hit rates per Day for a range of Cache Sizes. The cache algorithm is LRU, using a lookup key of the prefix and the Origin AS, with a validation re-use period of 36 hours.

Validation of BGP updates can apply to more than just the validity of the advertised address prefix. Using a signed attestation, where the address holder explicitly permits an Autonomous System to originate a BGP advertisement, the combination of the address prefix and originating AS can be validated. The extent to which caching the validation outcome of the combination of address prefix and origin AS in indicated in Figure 6. The number of prefixes that show short term instability in their origin AS is very low, so the validation cache outcomes when using a key of prefix plus origin AS are very similar. A cache of 1000 prefix + Origin AS validation entries can reduce the validation processing load by an average of 60% within a 24 hour period, while a cache size of 10,000 entries offers an average hit rate of 75%. This cache size, 10,000 entries, is some 5% of the total number of distinct entries in the BGP routing table as of September 2006.

The next step is to include consideration of the AS Path into the cache key. In this case the routing assertion that is being authenticated when validating the AS Path is that the BGP update was processed by each of the ASs in the AS Path in sequence, and that the AS Path has not been altered in any way. To replicate this load the validation cache key used the compressed AS path, where instances of AS prepending with duplicate AS number values in the path were removed The validation cache rate is indicated in Figure 7. In this case the validation cache is not as effective, and a cache of 10,000 entries will reduce the total validation load by some 30% to 50%. A potential explanation of this difference lies in the nature of the protocol operation of BGP. Withdrawal of a route may not propagate at a uniform rate through the network, and partial withdrawal information may generate transient routing states during the convergence period. These transient routing states share a common origin and AS number, but differ in the AS Path. The sequence of transient routing path states may be highly variable, which in turn causes a low level of cacheability of the path-based information.

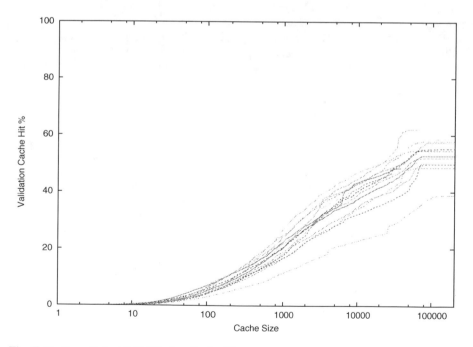

Fig. 7. Prefix + Path AS Validation Cache Hit rates per Day for a range of Cache Sizes. The cache algorithm is LRU, using a lookup key of the prefix and the Compressed AS Path, with a validation re-use period of 36 hours.

This data suggests that caching can offer the highest processing load reduction when the cache reflects the validation of the prefix and the origin AS, and that a cache size of between 1,000 to 10,000 entries can offer a reduction in processing overheads of between 60% to 80%.

4 Conclusion

BGP updates exhibit a distribution structure where a relatively small number of prefixes are the subject of a significant number of similar BGP updates, and these self-similar updates appear to be strongly clustered in both time and space.

A significant concern when examining proposals to modify BGP processing to include validating the security credential material supplied with BGP updates is the incremental processing load that may be imposed on BGP speakers.

If a BGP speaker is willing to cache a validation outcome for a period of up to 36 hours, and retain these outcomes using a cache size of 10,000 entries (or some 5% of the total number of distinct BGP table entries) with a LRU replacement cache management algorithm, then the cache is capable of performing at an average 75% hit rate for prefix origination. Similar parameters for prefix plus path validation indicate that a similar set of cache parameters, namely a cache size of 10,000 entries and 36 hours cache hold time, is capable of sustaining an average of a 30% to 50% cache hit rate.

This analysis indicates that the incremental load imposed by adding validation of credentials associated with BGP updates can be significantly mitigated by using caching of validation outcomes for subsequent reuse.

This work reflects a study of the BGP update traffic for a single peer. Future work on a study on self-similarity and cache parameters for multiple eBGP sessions, decoupled origination and path validation caching, and also the effects of delayed validation on the validation work load are logical extensions of this study.

References

1. Y. Rekhter, T.Li, S. Hares: A Border Gateway Protocol 4 (BGP-4), RFC 4271, Internet Engineering Task Force, January 2006.
2. Internet Architecture Board: Minutes of Meeting, January 1991 (online at http://www.iab.org/documents/iabmins/IABmins.1991-01-08.arch.html)
3. G. Huston: 2005 – BGP Updates, presentation to Global Routing Operations Working Group, IETF 65, March 2006 (online at http://www3.ietf.org/proceedings/06mar/slides/grow-3.pdf)
4. S. Kent, C. Lynn, and K. Seo: Secure border gateway protocol (s-bgp), IEEE Journal on Selected Areas in Communication, vol. 18, no. 4, 2000.
5. GNU Zebra. (online at http://www.zebra.org)

BGP Route Propagation Between Neighboring Domains

Renata Teixeira[1], Steve Uhlig[2], and Christophe Diot[3]

[1] Univ. Pierre et Marie Curie, LIP6-CNRS
renata.teixeira@lip6.fr
[2] Delft University of Technology
S.P.W.G.Uhlig@ewi.tudelft.nl
[3] Thomson Paris Lab
christophe.diot@thomson.net

Abstract. We propose a methodology to match detailed BGP updates from two neighboring Autonomous Systems (ASes). This methodology allows us to characterize route propagation and measure the route propagation time. We apply this methodology to two months of all BGP updates from Abilene and GEANT to perform the first thorough characterization of BGP route propagation between two neighbor ASes. Our results show that the propagation time of BGP routing changes is very different depending on the network that initiates the routing change. This difference is due to engineering and connectivity issues such as the number of prefixes per BGP session, the number of BGP sessions per router, and BGP timer configurations.

1 Introduction

Although Autonomous Systems (ASes) in the Internet are independent management entities, events such as equipment failures or router misconfigurations in one AS can trigger BGP routing changes that propagate to other ASes. During routing convergence, user traffic may encounter loops or loss of reachability. Besides these transient disruptions, BGP routing changes can also lead to persistent reachability problems, because there may be no route to the destination or because the new route may be incorrect (in case of a misconfiguration). A detailed characterization of the dynamics of BGP route propagation can help reduce the impact of routing changes in one AS on neighboring ASes and reduce convergence delay. Such a characterization can also play an important role in diagnosing the root cause of persistent problems. To troubleshoot the problem operators often need to pinpoint the AS responsible for the routing change.

In this paper, we make a major step toward understanding BGP route propagation between neighboring ASes. We introduce a methodology for correlating BGP routing changes in two neighboring networks based on BGP updates collected in each of the ASes. We use this methodology, together with two months of BGP updates from Abilene and GEANT (the research backbones in the U.S. and Europe, respectively) to analyze BGP route-propagation time. Our results show that although the types of BGP routing changes that propagate between these two networks are similar, the propagation time is significantly different depending on which of the two networks initiates the

S. Uhlig, K. Papagiannaki, and O. Bonaventure (Eds.): PAM 2007, LNCS 4427, pp. 11–21, 2007.

routing change. We show how this disparity is based on each network's design and engineering decisions, including factors such as the number of prefixes per BGP session, the number of BGP sessions per router, and the configuration of BGP timers.

This is the first time that BGP update measurements from every router in two neighboring ASes have been used to evaluate the impact of BGP routing changes on neighbors. Previous studies of BGP dynamics either analyzed BGP update messages from multiple routers in the same AS [1,2,3,4] or a single router in each of multiple ASes, as available from RouteViews or RIPE, combined with beacon updates [5,6,7]. Analyzing BGP updates in one AS can reveal how routing changes propagate within a single network, but does not shed light on how these changes affect neighboring domains. Studies of multiple ASes can characterize the BGP convergence process in the wide area, without shedding light on the effects of intra-AS topology and configuration. In this paper, we find that per-router BGP measurements and knowledge of the network design and configuration details are essential for understanding the factors that affect route-propagation time.

The remainder of the paper is structured as follows. Section 2 presents background on BGP routing between neighboring ASes. After presenting Abilene and GEANT in Section 3, we introduce our methodology for correlating BGP routing changes that propagate between neighboring ASes in Section 4. Section 5 quantifies the BGP routing changes that propagate between them and their propagation time. We end in Section 6 with a summary of our main findings and a discussion of their implications.

2   BGP in Neighboring ASes

Neighboring ASes connect in one or more physical locations, which we call *interconnection points*. Figure 1 illustrates two neighboring ASes X and Y, where $x1, x2, x3, x4$ and $y1, y2, y3, y4$ are routers in X and Y, respectively, and $p1, p2, p3, p4$ are destination prefixes. X and Y have two interconnection points $(y1, x3)$ and $(y2, x4)$.

Routers at interconnection points exchange reachability information to destination prefixes using external BGP (eBGP). We use the notation $P_{X \rightarrow Y}$ to refer to the set of prefixes that X announces to Y (even if Y is not using the route learned from X to that

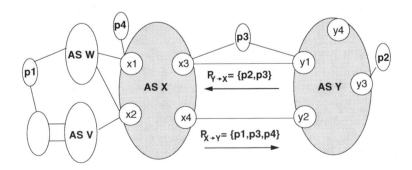

Fig. 1. Route propagation between two neighboring ASes X and Y

prefix). In the example, $P_{X \to Y} = \{p1, p3, p4\}$, even though Y might use the path to $p3$ it receives from elsewhere, instead of the route from X. BGP routing changes in X for prefixes that belong to $P_{X \to Y}$ may propagate to Y via the interconnection points.

A router can also learn a BGP route for a destination prefix from other routers in its own AS using internal BGP (iBGP). For example, router $y3$ learns the route to $p3$ from router $y1$. Each router selects the best route to reach this prefix using a multi-step decision process [8], which compares routes based on local policy preferences for path attributes (to a destination) such as AS-path length. Routes learned at all interconnection points to a neighboring AS often have the same AS path length, and other identical attributes. For example, X may learn equally-good routes to $p1$ at $x1$ and $x2$. We call each border router that receives a best route to reach a prefix p from eBGP an *egress router* for p, and the set containing all the egress routers for p as the *egress set* for p. For example, the egress set for $p1$ at AS X is composed of $x1$ and $x2$. Routers inside the AS break the tie among the routes learned from each router in the egress set by selecting the BGP route from the closest (in an intra-domain sense) egress router. This decision is commonly called *hot-potato* routing.

3   Abilene and GEANT

Abilene and GEANT are quite different networks. GEANT is an Internet service provider dedicated to academic institutions, whereas Abilene is a private academic network that is not connected to the commercial Internet. As we will see, these differences allow us to illustrate the impact of engineering decisions and network connectivity on route propagation.

3.1   Inter-connectivity

Abilene and GEANT have a peer relationship to exchange traffic between their respective customers. Since Abilene is not an Internet provider, all networks that connect to Abilene must have a separate connection to the Internet, by which they can also reach GEANT's customers. GEANT, on the other hand, has six connections to the commercial Internet. GEANT routers have BGP tables with approximately $170,000$ destination prefixes, whereas the BGP tables for Abilene routers have slightly under $10,000$ prefixes.

Research and academic institutions in Europe connect to GEANT through national or regional research networks. Some of these national academic networks have their own connectivity to commercial ISPs. On the other hand, Abilene connects directly to individual institutions. Because of its connection policy, GEANT has many more opportunities for route aggregation, which explains why Abilene announces to GEANT twice as many BGP prefixes than GEANT to Abilene ($|P_{A \to G}| = 5,770$ whereas $|P_{G \to A}| = 2,200$).

Abilene and GEANT have two peering links: between Washington DC (WA) and Frankfurt (DE2), and between New York (NY) and Amsterdam (NL). Abilene and GEANT announce BGP routes with equal AS-path length in both peering locations, and use the same local preference value in both locations as well. Consequently, each router selects between the two interconnection points using hot-potato routing. Neither of the two networks use BGP's Multi-Exit Discriminator (MED) attribute.

3.2 Measurement Infrastructure

Both Abilene and GEANT use Juniper routers running a full-mesh of iBGP sessions. BGP monitors in both networks are NTP synchronized. However, their measurement infrastructure differs significantly. Abilene has one Zebra BGP monitor per PoP. Given that there is only one router per PoP in Abilene, each monitor establishes an iBGP session as a *client* of the router and collect periodic table dumps as well as all BGP messages reporting changes to the best route to each prefix. The union of BGP messages from all routers gives a global view of each router's choice of best routes for each prefix. GEANT uses a single Zebra BGP monitor that participates in the iBGP full mesh. This monitor is configured as an iBGP peer and thus only receives BGP messages reporting routes learned from eBGP. It does not receive BGP update messages triggered by internal routing changes.

4 Measurement Methodology

This section describes our methodology to correlate BGP update measurements in neighboring ASes. For these correlated BGP changes, we also compute the time it takes until the BGP change in X causes a change in Y, and vice-versa.

4.1 Classification of BGP Changes

First, we classify BGP routing changes from the vantage point of each AS according to the three categories described below, which are inspired from [3]. The main distinction between our work and [3] is that they evaluate the impact of routing changes at X on X's traffic, whereas we classify the BGP routing changes in an AS X that propagate to a neighboring AS Y[1].

Prefix down in X. When X looses connectivity to a prefix $p \in P_{X \to Y}$, each border router sends a message reporting the withdrawal of p. This withdrawal may impact Y in two different ways: routers in Y also withdraw p or shift to another route that does not use X.

Prefix up in X. Similarly, when X gains connectivity to a prefix $p \in P_{X \to Y}$, each border router x sends an announcement of p. Routers in Y may experience a prefix up as well, in the case Y did not have a route to p before receiving the update message; or an egress-set change to use the route from X.

Egress-set change in X. We define an *egress-set change* as a BGP event that changes the composition of the egress set for a given prefix. Routers in X can still reach p, but decide to change routes because the previous route was withdrawn or a new (better) route came up. There are three different types of egress-set changes in X: a change to a worse, equivalent, or better route. For example, suppose that the link between AS W and $p1$ fails in Figure 1. X would then replace this route with the one through AS V, which is worse than the previous route because it has an AS-path length of two, instead of a length of one via W. This change would not trigger an egress-set change in Y, because even though the new route via X is worse, Y does not have a better alternative.

---

[1] Although intra-domain routing changes can also impact neighboring ASes because of hot-potato routing [1] or cold-potato routing [8], we do not consider these type of changes here.

In the case that Y has another route to $p1$ that is better than the new route via X, then Y would change routes to $p1$.

The collection of BGP update messages from all routers in an AS contains a lot of redundancy. Indeed, multiple routers report the same routing change, and a single router may also send multiple messages for the same prefix in a very short period of time because of path exploration [5]. The main classification challenge is therefore to extract one instance of each BGP routing change from all BGP update messages. We extract BGP routing changes using the methodology described in [3]. For each prefix, we group all BGP routing changes that happen close in time [1,3]. For the results presented in this paper, we select a 70-second threshold to eliminate redundant BGP update messages (approximately 75% of the BGP updates, which is consistent with [3]). We use the timestamp of the first BGP update in the group of updates that leads to a BGP routing change as the timestamp for the change.

4.2   Correlating BGP Routing Changes

Given a time series of labeled BGP routing changes and a time window T, we determine which of the BGP routing changes at Abilene propagate to GEANT, and vice-versa. We call an AS X the *source* of a change, if the routing change happens first at X, and then propagates to Y (which we call the *destination*). We develop a routing correlation algorithm that proceeds in two steps:

Selection of relevant BGP routing changes. We measure $P_{A \rightarrow G}$ using BGP table snapshots and BGP messages collected at GEANT. Since we want $P_{A \rightarrow G}$ to contain *any* prefix that might be announced by Abilene to GEANT during our analysis, we search for any destination prefix that has at least one BGP message with next-hop AS equal to Abilene's. Similarly, we search Abilene's BGP messages to extract $P_{G \rightarrow A}$.

If $P_{A \rightarrow G} \cap P_{G \rightarrow A} \neq \emptyset$, then the causal relationship between BGP routing changes to a destination prefix $p \in P_{A \rightarrow G} \cap P_{G \rightarrow A}$ is not clear. In fact, each AS should use its direct route to p most of the time, except for transient periods of failures. Therefore, we exclude all prefixes in $P_{A \rightarrow G} \cap P_{G \rightarrow A}$ from our analysis to focus on the set of *distinct* destination prefixes that Abilene announces to GEANT, and vice-versa.

Matching related BGP routing changes. Our algorithm first reads the stream of BGP routing changes of Y and creates a list of time-ordered changes per destination prefix. Then, we identify whether each BGP routing change of X triggered a change in Y. For each BGP routing change for a prefix p in X of type c at time t, we search the list

Table 1. Compatibility of BGP routing changes at neighboring ASes

Type at source AS	Type at destination AS
prefix down	prefix down
	egress-set change
prefix up	prefix up
	egress-set change
egress-set change	egress-set change

of changes to p in increasing time order. We say that a change in X triggered another in Y of type c' at time t', if $t \leq t' \leq t + T(p)$ and c' is *compatible* with c. We define compatibility as follows. Two routes are *compatible* if the type of BGP routing change at the source and destination ASes falls into one of the categories in Table 1. This algorithm returns the list of BGP routing changes of X, where each change is annotated with the corresponding change in Y or a null value.

Given the frequent churn of BGP messages caused by events at several locations in the Internet, any heuristic to match BGP routing changes at neighboring ASes has the risk of mistakenly correlating two BGP routing changes that did not propagate between the neighbors in question. Take the example in Figure 1 and suppose that Y uses another neighbor (not shown in the figure) to route to $p1$. A failure at $p1$'s network could cause a prefix down both at X and Y, even though the BGP routing change did not propagate from X to Y. Our algorithm would mistakenly correlate these routing changes. Although we leave a detailed study of these false matches for future work, we include some tests in our algorithm to reduce the likelihood of these false matches:

- **Selection of the prefixes to consider**. We search BGP tables from both networks to determine $P_{A \to G}$ and $P_{G \to A}$ and remove prefixes in the intersection, which could lead to false matches.
- **Classification of BGP routing changes**. We ensure that only compatible routing changes are correlated.
- **Selection of time window T**. The time window guarantees that events that happen too far apart do not get correlated. We set this time window to the worst-case propagation time between the two neighbors. By using the worst-case propagation time, we guarantee to find all truly correlated BGP routing changes while limiting the number of false matches. The next section explains the procedure to find the worst-case propagation time from network configuration data and BGP tables.

4.3   Worst-Case Time Propagation

Our correlation algorithm searches for BGP routing changes that happen *close* in time at both networks, where close means within a time window T. Since routing configurations are different in Abilene and GEANT, we choose a different time window $T_{A,G}$ from Abilene to GEANT and $T_{G,A}$ from GEANT to Abilene. We define the time window as the worst-case BGP propagation time among all the interconnection links.

The propagation of a BGP message is influenced by iBGP (to transfer the message from the egress router to the interconnection point), and eBGP (to transfer the message between the interconnection routers). Juniper routers use an "out-delay" timer to avoid sending updates too often. eBGP sessions may also apply the "route-flap damping" mechanism upon the reception of BGP messages coming from an external neighbor. Another important factor is router load. A measurement study of BGP "pass-through" times [9] showed that the number of prefixes advertised in a BGP session and the number of BGP peers are key contributors to router load. Propagation time also depends on other properties such as network propagation delay and route reflector hierarchy, but neither are relevant here.

The propagation time from GEANT to Abilene has two main components: an out-delay and a load-related delay. GEANT sets the out-delay at the interconnection sessions

10 seconds at (NL,NY) and 30 seconds at (DE2,WA). The worst-case scenario for the transfer delay would be a reset of one of the sessions with GEANT's providers. In this case, the transfer of the 170, 000 routes to one iBGP neighbor should take around 3 minutes [1]. If the router issuing the updates is CPU bound, which is usually the case when it has to treat a large number of updates, then it will send updates to each neighbor sequentially. If the BGP monitor was the last iBGP neighbor to receive the updates, then it would only receive an update reporting a change after all the other 21 neighbors (i.e., 21×3 minutes after the interconnection point received the change). Therefore, we bound the propagation time of events from GEANT to Abilene with a one-hour time window.

The time window for the propagation of routing events from Abilene to GEANT is mainly determined by route-flap damping imposed by GEANT at the reception of updates from Abilene. GEANT sets the maximum delay introduced by route-flap damping mechanism according to the RIPE recommendations [10] (i.e., 30, 45, and 60 minutes for short, medium, and long prefixes). Abilene does not set the out-delay timers and there is little load-related delay (Abilene's largest BGP session is with GEANT, and it only has 2, 200 prefixes). Therefore, we use an adaptive time window that depends on the prefix length: $T_{A,G}$ is 1820 seconds, if prefix is shorter than /22; 2720 seconds, if prefix is /22 or /23; and 3620 seconds, if prefix is longer than /24.

5   Analysis of Route Propagation

We now analyze each pair of BGP routing changes from Abilene to GEANT, and vice-versa, correlated according to the methodology described in Section 4. First, we characterize which kinds of BGP routing changes are more frequent and therefore have a more significant impact between Abilene and GEANT. Then, we quantify the route propagation time.

5.1   Classification of Propagated Routes

Table 2 presents the number of *BGP routing changes* per type as defined in Section 4.1. The first half of the table presents the number of BGP routing changes at the source AS. The second half quantifies the *impact* of these changes on the destination AS.

Table 2. BGP routing changes that propagate between Abilene and GEANT

BGP routing change			Impact		
Type	Abilene	GEANT	Type	Abilene to GEANT	GEANT to Abilene
prefix down	19, 109	4, 318	prefix down	5, 496	1, 506
			egress-set change	3, 636	94
prefix up	22, 262	6, 214	prefix up	7, 467	2, 558
			egress-set change	4, 803	316
egress-set change	6, 925	3, 591	egress-set change	82	0
total	48, 296	14, 123	total impact	21, 484	4, 474

Abilene experienced $48,296$ BGP routing changes that could potentially impact GEANT during the measurement period, whereas GEANT only experienced $14,123$ BGP routing changes that could impact Abilene. This difference is explained by a combination of two factors: (i) the number of prefixes in $P_{A \to G}$ is more than twice the number in $P_{G \to A}$, and (ii) Abilene does not apply any delay to filter BGP messages, which leads to a higher number of BGP routing changes. Both sets of results show that prefix up and down events dominate the routing changes of each network (these events represent 85.7% of events at Abilene and 74.6% at GEANT).

The first line of Table 2 shows that there were $19,109$ prefix-down events at Abilene that could impact GEANT, but that less than half of those $(9,132)$ actually triggered a BGP routing change at GEANT. One reason is that the route-flap damping mechanism applied by GEANT filters many of these events. Another reason is that GEANT can also reach most prefixes that it learns from Abilene using its own connection to the commercial Internet. If GEANT is not using the route via Abilene, then the loss of reachability in Abilene does not impact GEANT.

Given the limited number of alternative paths that Abilene has to reach the prefixes announced by GEANT and vice-versa, most egress-set changes at the source AS have no impact at the destination. In particular, Abilene has almost no alternative to reach the prefixes announced by GEANT ($P_{G \to A}$). Therefore, even when a prefix goes down or if GEANT changes to a worse route, Abilene routers have no alternative but to lose connectivity to the prefix or still select the route they learn from GEANT, respectively. There are only 94 instances in which a prefix down at GEANT caused Abilene to replace its egress set to the prefix. We have verified some of these events manually and observed that Abilene and GEANT have some common peers (mainly research and educational networks in Latin America, Asia-Pacific region, and Africa). Some prefixes are multi-homed to GEANT and to one of these other peers. Abilene uses either one of the routes to reach these prefixes, and events at GEANT cannot impact Abilene when it is using the route via the other peer. This behavior explains why only $1,600$ out of the $4,318$ prefix down at GEANT trigger a change in Abilene.

Table 2 illustrates the types of routing changes that propagate between these two academic peers. We expect the types of routing changes to vary substantially for different pairs of neighboring ASes, because of their relationship, the number of connections to their neighbors, and their location in the Internet hierarchy. GEANT and Abilene experience mostly gain or loss of reachability (or, "prefix up" and "prefix down"). Both networks are fairly small and are close to the edge networks, which implies that they are closer to the network that originates the BGP routing change and that there is less aggregation of prefixes. These results are in sharp contrast with the 6.0% of loss/gain of reachability measured at a tier-1 ISP network [3]. The majority of events at the tier-1 network were distant/transient disruptions, which we classify as egress-set changes.

5.2 Propagation Time

We estimate the propagation time of a routing change to a prefix p between Abilene and GEANT by comparing the time BGP monitors at each network receive the *first* BGP message that reports the routing change to p. We compute the propagation time

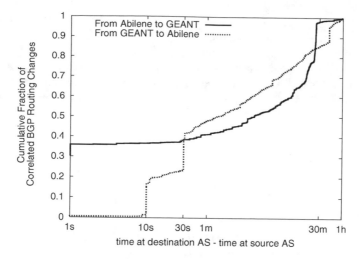

Fig. 2. Propagation time of routing changes

from a source to a destination AS as the difference between the time of the BGP routing change at the destination and the time at the source. Figure 2 presents the cumulative distribution of the propagation time of all correlated BGP routing changes. (Note that the x-axis is in log scale.)

Although the propagation time in both directions is less than one minute for approximately half of the correlated BGP routing changes, the shapes of the curves are strikingly different. Abilene does not use out delay, therefore over 35% of GEANT BGP routing changes triggered by Abilene happen within the first second in GEANT. The linear increase of the propagation time is an effect of a combination of the route-flap damping mechanism to enter GEANT, and of other load-related variations (as examined in [9]). The propagation time reaches a plateau at around 30 minutes, which we suspect is due to false matches (there are less than 2% of correlated events with more than 30 minutes propagation time).

On the other hand, the analysis of the propagation time from GEANT to Abilene shows that almost all BGP routing changes from GEANT take at least 10 seconds to reach Abilene. Indeed, the distribution of propagation time has two distinguishable steps at 10 and 30 seconds, which correspond to the out delay of 10 seconds imposed by GEANT at the NL router and of 30 seconds at DE2. The propagation time of almost half of BGP routing changes from GEANT to Abilene is determined by these timers. The slow increasing slope is due to the interaction of a number of factors: TCP behavior at the BGP session, the CPU load at the border router, and the number of BGP messages triggered by each BGP routing change.

We examined the small steps in the distribution from GEANT to Abilene that appear around 3, 7, and 40 minutes propagation times. We found that all of these steps correspond to BGP session resets. A session reset triggers a large number of BGP messages. All these messages reach the neighboring AS at approximately the same time, and consequently have similar propagation times. We conjecture that the few BGP routing changes with propagation time over 30 minutes are due to the load in the GEANT router that first experiences the change (as discussed in Section 4). If the BGP monitor is among the first peers to be notified of a large session reset and the interconnection points to Abilene are among the last ones, we expect time lags even larger than 30 minutes. For instance, the sharp increase in the time propagation distribution around 40 minutes from GEANT to Abilene happens because of the re-establishment of a session with one of the providers. Certainly, 40 minutes of propagation time between neighboring networks is extremely large, but also rare. This example illustrates the importance of taking into account the router load as a factor of propagation time. Events such as session resets or hot-potato routing changes can trigger thousands of routes to change at the same time [1,3], and hence substantially increase the load in the router.

6   Conclusion

This paper shows that BGP route propagation is most sensitive to engineering and connectivity of the networks it traverses. The propagation of BGP routing changes between neighboring ASes can sometimes take more than 30 minutes. The longest propagation times from Abilene to GEANT are due to route-flap damping. From GEANT to Abilene, the highest propagation times are caused by the load of the router where routes are processed. GEANT has BGP sessions where it learns more than 150,000 prefixes from a neighboring AS. A reset of any of these sessions would generate a prohibitively large number of BGP updates that would in turn impact the router's load. Note that any AS that has a provider can experience a similar phenomenon, because ASes learn full BGP tables in the session with their providers. The number of prefixes exchanged in each BGP session and the number of BGP sessions per router are important factors that impact router load.

Acknowledgments

We would like to thank Abilene and GEANT for making their measurement data publicly available to the research community. We also thank Richard Gass from Intel and Nicolas Simar from DANTE for helping us understand GEANT's data and engineering. We are grateful to Jim Kurose, Olaf Maennel, Jennifer Rexford, and Augustin Soule for their insightful comments on this work.

References

1. R. Teixeira, A. Shaikh, T. Griffin, and J. Rexford, "Dynamics of Hot-Potato Routing in IP Networks," in *Proc. ACM SIGMETRICS*, June 2004.
2. S. Agarwal, C.-N. Chuah, S. Bhattacharyya, and C. Diot, "Impact of BGP Dynamics on Intra-Domain Traffic," in *Proc. ACM SIGMETRICS*, June 2004.
3. J. Wu, Z. Mao, J. Rexford, and J. Wang, "Finding a needle in a haystack: pinpointing significant BGP routing changes in an IP network," in *Proc. USENIX Symposium on Networked Systems Design and Implementation*, May 2005.
4. D. Pei and J. V. D. Merwe, "BGP convergence in MPLS VPNs," in *Proc. Internet Measurement Conference*, 2006.

5. C. Labovitz, A. Ahuja, A. Bose, and F. Jahanian, "Delayed Internet Routing Convergence," *IEEE/ACM Trans. Networking*, vol. 9, pp. 293–306, June 2001.
6. Z. M. Mao, R. Govindan, G. Varghese, and R. Katz, "Route Flap Damping Exacerbates Internet Routing Convergence," in *Proc. ACM SIGCOMM*, August 2002.
7. R. Oliveira, B. Zhang, D. Pei, R. Izhak-Ratzin, and L. Zhang, "Quantifying path exploration in the internet," in *Proc. Internet Measurement Conference*, 2006.
8. S. Halabi and D. McPherson, *Internet Routing Architectures*. Cisco Press, second ed., 2001.
9. A. Feldman, H. Kong, O. Maennel, and A. Tudor, "Measuring BGP pass-through times," in *Proc. of Passive and Active Measurement Workshop*, pp. 267–277, 2004.
10. C. Panigl, J. Schmitz, P. Smith, and C. Vistoli, "Recommendations for coordinated route-flap damping parameters," October 2001. `http://www.ripe.net/ripe/docs/routeflapdamping.html`

Detectability of Traffic Anomalies in Two Adjacent Networks

Augustin Soule[1], Haakon Ringberg[2], Fernando Silveira[3], Jennifer Rexford[2], and Christophe Diot[1]

[1] Thomson Research
[2] Princeton University
[3] Federal University of Rio de Janeiro

Abstract. Anomaly detection remains a poorly understood area where visual inspection and manual analysis play a significant role in the effectiveness of the detection technique. We observe traffic anomalies in two adjacent networks, namely GEANT and Abilene, in order to determine what parameters impact the detectability and the characteristics of anomalies. We correlate three weeks of traffic and routing data from both networks and apply Kalman filtering to detect anomalies that transit between the two networks. We show that differences in the monitoring infrastructure, network engineering practices, and anomaly-detection parameters have a large impact on which anomaly detectability. Through a case study of three specific anomalies, we illustrate the influence of the traffic mix, IP address anonymization, detection methodology, and packet sampling on the detectability of traffic anomalies.

1 Introduction

Identifying anomalous Internet traffic, such as malicious attacks, flash crowds, or traffic shifts, is a difficult and important challenge for Internet Service Providers (ISPs). In the past few years, researchers have introduced a promising new way to detect anomalies. Rather than scrutinizing the traffic on each link independently, the traffic is summarized in a link or traffic matrices and analyzed on all links simultaneously. Then, anomalies are detected by applying statistical analysis techniques to the matrices. This is known as "network-wide" anomaly-detection. This approach is very effective at detecting anomalies that are spread over multiple links, such as distributed attacks or traffic shifts caused by routing changes[1,2].

Despite promising initial results, we still understand very little about network-wide anomaly detection methods. Relatively few papers have been published, and these studies unfortunately (1) do not describe the calibration of the methodology very accurately and (2) do not use the same measurement data sets. In addition, identifying and classifying *all* anomalies in a given traffic trace (in order to get a *ground truth* to which to compare the outcome of the detection methods) is extremely difficult, if not impossible on a large data set. The number of anomalies detected depends on many parameters that have not been studied systematically. In earlier papers, "manual tweaking" and "visual inspection" play an important role in the success of the anomaly-detection techniques.

S. Uhlig, K. Papagiannaki, and O. Bonaventure (Eds.): PAM 2007, LNCS 4427, pp. 22–31, 2007.

Therefore, network administrators cannot readily apply these network-wide anomaly-detection techniques "out of the box" or easily tune them for effective use in their networks. It will not be possible to use these methods in an operational network until we understand how anomaly "detectability" is influenced by network design, monitoring infrastructure, and anomaly-detection technique.

In this paper, we take a first step in this direction by studying traffic anomalies simultaneously in two backbone networks—GEANT and Abilene[1], focusing on anomalies that cross both networks. We analyze three weeks of time-synchronized traffic and routing traces for the two networks. Note that the goal of this work is not to answer all questions and explain every observation. We are far from being able to do so for reasons explained earlier. Our ambition is to identify problems and issues that need to be addressed before thinking of unsupervised and automatic anomaly detection in operational environment.

We summarize the traffic (for each time interval) in four entropy *Link Matrices*, each matrix corresponding to a given flow feature (i.e. source and destination IP address and source and destination port). To detect anomalies, we use the Kalman-filtering method introduced in [3]. The measurement data sets and detection methodology will be described in more details in the next sections. Note that for the purpose of this work, the detection method is not critical as long as it detects real anomalies with a low false positive ratio.

We use BGP routing information to identify the subset of the traffic that traverses both networks, and we perform anomaly detection on these reduced data sets, as well as on the full link matrices. Surprisingly, we find that many anomalies are detected in one network and not in the other. We show that it can be due to (i) the difference in monitored traffic sampling rate, (ii) the anonymization of the IP addresses, (iii) the calibration of the detection method (i.e., value of the detection threshold), or simply because of (iv) the traffic mix on the link where the anomaly is detected. To illustrate these claims, we analyze three specific anomalies where each of the potential causes listed above is involved in a missed anomaly detection in one of the two networks.

The paper is organized as follows. Section 2 describes our measurement data and formalism. Section 3 presents the anomaly detection methodology. We start the discussion of results (section 4) by general observations about the anomalies that are detected and missed in each network and the factors that impacts the detectability. Section 5 illustrates the general discussion with three specific anomalies where previously identified factors are indeed causing an anomaly to be missed. We discuss research challenges and future research directions in section 6.

2   Measurement Data

2.1   Collecting the Traffic and Routing Measurement

The data used in this paper has been collected in two academic networks, GEANT and Abilene. Abilene provides connectivity to research and academic networks in the US.

---
[1] www.geant.net and abilene.internet2.edu

It has 11 points of presence (PoPs) and 198 incoming links. One peculiarity of this network is that each Abilene customer must have a separate connection to the Internet since Abilene does not connect to the Internet. Abilene is very interesting from an anomaly-detection standpoint, as it is mostly used for experimental academic traffic. GEANT is the European Research Network. It interconnects national research networks, rather than directly connecting research institutions. GEANT is composed of 22 PoPs and 99 incoming links. It is connected to the Internet and provides transit service to its customers. During the time our data was collected, Abilene and GEANT were peering at two locations: between Washington DC and Frankfurt through an OC48 link and between New York and Amsterdam through a virtual LAN. Note that the traffic from multiple Autonomous Systems (ASes) is mixed in this VLAN. Therefore, we can not isolate the traffic going from Abilene to GEANT, and vice-versa, from the recorded data.

Both networks collect routing and sampled traffic statistics. Abilene collects routing information through Zebra BGP monitors connected to the routers. GEANT has one single Zebra BGP monitor which is part of the iBGP mesh. In both cases, the BGP monitors record all BGP updates. The flow statistics are recorded on each router using Juniper's *J-Flow* tool. GEANT routers' record one out of every 1000 packets and the flow information is exported to the Network Operation Center (NOC) every fifteen minutes. In Abilene packets are sampled at 1 out of 100, and flow information is exported every five minutes. In Abilene, the last 11 bits of each IP address are set to zero, preventing the identification of the source or destination host.

Merging all four datasets has been a serious challenge. We could only identify a period of 20 consecutive days between November 10 and November 30, 2005, where all datasets were complete in both networks. Routers and monitors are synchronized using NTP and each measurement record is labeled with a timestamp. The datasets are collected at different geographic locations. We can not guarantee that the clocks are perfectly synchronized. But the time granularity of the flow statistics is much larger than the NTP error and thus time synchronization is not an issue.

2.2   Aggregating the Traffic into a Link Matrix

We are primarily interested in anomalies for which the traffic transits from Abilene to GEANT and GEANT to Abilene. However, direct observation of anomalies on the peering links is not possible because of the presence of the VLAN. Therefore, we opted for network-wide detection in each network, as described in [1,3,4]. The traditional Traffic Matrix defined in these papers is sensitive to routing changes [5]. Given that we need to match traffic between two networks, routing errors could bias our observations. Therefore, we chose a traffic formalism which is insensitive to internal routing changes, i.e. incoming link matrix, all the incoming link traffic time series combined in a single matrix. Routing information is then used to identify the subset of flows that go from Abilene to GEANT, and vice-versa. The anonymization in Abilene did not bias this step as we did not identify any prefixes longer than 21 bits from GEANT to Abilene. Therefore, we use four different sources of data: link matrix from Abilene (A) and GEANT (G), and link matrices made of the flows that go from Abilene to GEANT and GEANT to Abilene. These two link matrices are noted respectively $A2G$ and $G2A$. Note that any anomaly detected in $A2G$ or $G2A$ should also be detected in G and A respectively.

2.3 Detecting Anomalies

The flow statistics are represented by time series of entropy values computed on four IP header fields, namely source and destination IP addresses and ports, as defined in [1]. The entropy measures how a distribution is spread over the range of values. We use the classic entropy equation for each feature f: $X_f(t) = -\sum_i p_f(i,t) \log(p_f(i,t))$ where $p_f(i,t)$ is the proportion of packets containing the feature value i during the time interval t. The entropy is low when the distribution is concentrated on a few values, and high when each value is equally probable. Lakhina established in [1] that a significant variation in entropy is an effective way to identify the presence of an anomaly in the data set. The four features entropy based detection also helps identifying the cause of the anomaly.

Network-wide anomaly detection is performed using the Kalman method applied independently on each of the four features. This method was introduced in [3] and in [2]. These papers can be read for details on the detection method. In short, the Kalman filter extracts the predictable part of the traffic time series according to a predefined model. The difference between the prediction of the model and the observed value is defined as the "innovation" of the time series. Anomalies are defined as a significantly large difference between the predicted value and the observation. This significant change is identify as abnormal whenever the absolute value of the innovation exceed T times the variance (σ) of the innovation.

However, network-wide anomaly detection returns a list of time bins were an anomaly should be present based on the interpretation of entropy feature variation (together with the links where the anomalous time bin has been detected). *Network-wide anomaly detection tools does not detect anomalous traffic.* To identify the anomalous traffic we perform a post-mortem analysis of all anomalous time bins observed in both networks. We compare the traffic in the anomalous time bin to the one in the time interval that just precedes it and look for some significant change in the traffic that could explain a certain combination of entropy variation. An entropy decreases correspond to a concentration of the feature distribution and an entropy increase denotes a dispersion of the feature distribution. It is easy to identify the traffic that caused the entropy to decrease as it is, most of the time, due to a flow that increases its traffic. On the other hand, it is very difficult to identify the traffic corresponding to an entropy increase as what we are now looking for is a dispersion of traffic. Therefore, anomaly classification is easier when one of the feature exhibits an entropy decrease. We also try to aggregate the anomalies that are detected in consecutive time bins. Aggregation is performed by matching feature entropy values, links where the anomaly has been detected, and the IP flows carrying the anomalous traffic. Once aggregation has been performed in each network, we match $A2G$ anomalies to G anomalies (and $G2A$ to A). This labeled data and the associated methods are available upon request.

3 General Observations

The goal of this section is to identify what factors impact the detectability of traffic anomalies. Table 1 summarize the number of anomalies detected in each data set. This table also gives the number of anomalies that are found in two data sets simultaneously (i.e. $A2G$ and G, $G2A$ and A). The threshold is similar in both networks and equal to 10 times σ with $\sigma = \{\sigma_1 \cdots \sigma_n\}$ the variance of the innovations. We chose this value because it resulted in zero false positive in [2].

Table 1. Number of anomalies observed between Abilene and GEANT for a threshold of 10σ. 2005/11/10 to 2005/11/30.

Anomalies detected in					
A	G	$A2G$	$G2A$	$A2G \cap G$	$G2A \cap A$
78	14	58	10	5	3

78 anomalies are detected in Abilene and only 14 in GEANT. It is difficult to explain such a result, which only advocates for using a different threshold in both networks. We will come back on this issue below. GEANT is a larger network, but its sampling rate is lower. We conjecture that the very low sampling rate accounts for most differences in anomaly detection.

More anomalies are found in A going to G than in A coming from G. This phenomenon is more pronounced from A to G than from G to A, most probably because of sampling. This simply highlight the impact of the detection technique and the traffic data formalism on the detectability of anomalies. $A2G$ and $G2A$ are reduced data sets. They correspond to the subset of traffic captured in the origin network that is destined to the adjacent network. We conjecture that Kalman can extract anomalous behavior more easily in the reduced dataset.

This reduction of the data helps the Kalman method to detect anomalies. This is an interesting observation as it proves that both the method and the data set formalism impact the anomaly detectability.

We detected 58 anomalies in $A2G$ and only 5 (i.e. 9%) of these anomalies were detected in G. Similarly, 10 anomalies are detected in $G2A$ and only 3 of them are also detected in A (i.e., 33%). We suspect that the most probable explanation is the sampling rate in G, which is 10 times lower than in Abilene. The impact of sampling can also be observed on the number of $A2G$ anomalies detected in G, i.e. 12%. Moreover we expect that sampling affects differently the detectability of anomalies based on their nature. Thus the impact of the sampling rate on the anomaly detectability is not easy to evaluate. Recently a paper [6] studied the impact of traffic sampling on the detectability of the Blaster worm event using an entropy-based detection method. The paper shows that the worm is almost undetectable with a sampling rate of 1 out of 1000. In the same mindset, a theoretical result [7] shows that even a task as simple as ranking the flows according to the their size using sampled traffic requires a sampling rate greater or equal to 10%. We are far from these values in GEANT. However, not all anomalies in $G2A$ are also detected in A, which means that the sampling rate alone does not explain why we do not detect the same anomalies in both networks.

Anonymization could impact anomaly detection in Abilene. However, it is not clear whether the anonymization of the last eleven bits of the IP addresses reduces or increases the number of anomaly detected. However, we conjecture that anonymization will most probably change the way an anomaly is classified, by transforming the entropy dispersion of the IP address in an entropy concentration (showing one single IP address instead of multiple ones with the same prefix).

It is difficult to compare anomalies in the two networks with the same detection threshold. In the experiment below, we have chosen the threshold in G to be such that we obtain approximately the same number of anomalies in G and A. This threshold value is 5σ (when Abilene's threshold remains at 10σ). Table 2 show the number of anomalies in all data sets with this new threshold in G.

Table 2. Number of anomalies observed in GEANT for a 5σ detection thresholds. 2005/11/10 to 2005/11/30.

T_h in GEANT	anomalies detected in			
	G	$G2A$	$A2G \bigcap G$	$G2A \bigcap A$
Low	84	89	23	17

As expected, we now detect 84 anomalies in GEANT instead of 14 with a 10σ threshold which is five times more than with the 10σ threshold of table 1. Not surprisingly, the number of anomalies found in $A2G$ and G and in $G2A$ and A is also around 5 times more.

The 89 anomalies in $G2A$ are also easy to explain. Remember that $G2A$ is a different Link Matrix than G (in fact a subset the traffic contained in G). A higher number of anomalies in $G2A$ than in G confirms the impact of the detection method and of the data formalism on anomaly detectability, which has been discussed earlier in this section.

To summarize our observations, we have shown that the detection methodology, the data formalism, and the sampling rate do impact the number of anomalies that can be detected. We have seen that NOCs can play with the detection threshold to increase or decrease the number of anomalies detected. We suspect that IP address anonymization has a limited impact on anomaly detectability.

4   Case Study

The following three anomalies illustrate how the factors identified in the previous section can impact anomaly detection.

4.1   Impact of Sampling and Detection Threshold

This first anomaly is detected in the traffic flowing from Abilene to GEANT but is undetected in GEANT. It is an attacks against a SSH server that originated in Abilene on November 16th between 01:00 and 01:30 GMT. A host in the university of Philadelphia starts scanning the network for vulnerable servers. It finds a reachable SSH server in Italy (at 01:15 GMT). Then the attacker tries to gain access to this server by flooding it with SSH packets.

In figure 1, we show the entropy of the four features (source IP, destination IP, source Port and destination Port) as seen by Abilene in its New York router 1(b) and in GEANT on its peering link to Abilene 1(a). The vertical line indicates the time at which the

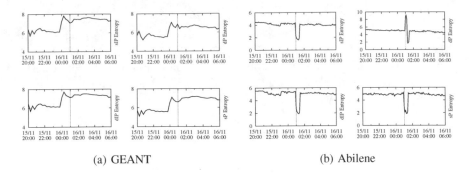

(a) GEANT   (b) Abilene

Fig. 1. Entropy of the four features during the SSH attack as seen in GEANT and Abilene

anomaly was detected in Abilene. The entropy plots on Abilene (fig. 1(b)) show that all the entropy values except the destination port decrease as expected in the case of port scans. The destination ports entropy increase shows that the attacker is doing a port scan on a few set of machines. At 01:15 GMT the attacker has found its victim and now targets its attack on a single port of the victim. As a consequence, the destination port entropy decreases.

This event can been seen on GEANT (fig. 1(a)) as a small increase followed by a small decrease on the destination port entropy. But the amplitude of the change is too small to be detected as an anomaly. Indeed, the total amount of anomalous traffic in Abilene is 84 000 packets. In GEANT, only 9 000 packets are sampled for the same traffic (i.e. approximately one tenth).

This anomaly being observed in GEANT, it is interesting to discuss whether a lower threshold in GEANT would have made this anomaly detectable. The figure 2 represents the time series of the Kalman innovation divided by its variance for the four features on each network. As seen in the figure 2(b) any threshold lower than 12σ in Abilene will detect this anomaly. But inside GEANT the threshold need to be set to at most 1σ to be able to detect this event.

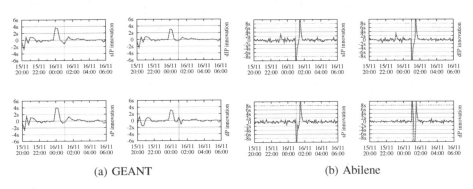

(a) GEANT   (b) Abilene

Fig. 2. Kalman innovation of the four features during the SSH attack as observed in GEANT and Abilene

4.2   Impact of the Traffic Mix

This non malicious anomaly was detected in GEANT only on November 16th at 10:00 GMT. It is characterized by a small number of SSH flows transferring a large amount of data between two hosts, one in the UK, and the other near New York. This transfer is performed over port 22, so we suspect these flows use SFTP. The four features observed in GEANT are shown figure 3(a). These features are the one we would expect in a such case. As in the case of a large file transfer between two hosts using a known application, the entropy of all four features should decrease as observed in figure 3(a). However, this anomaly was not detected in the Abilene despite a higher sampling rate and also despite that around 30,000 packets belonging to the anomaly were sampled on Abilene. The entropy of the features observed at this time on Abilene are shown figure 3(b). The reason why this anomaly goes undetected on Abilene is that at the same time, the entropy captures a concentration on port 80 due to on on-going massive HTTP transfer (220 000 packets in the anomalous time bin). Our anomalous file transfer is not big enough to significantly impact the entropy.

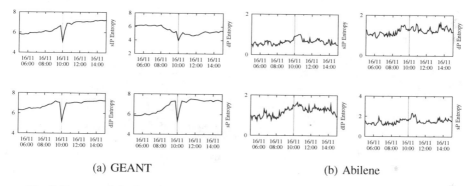

(a) GEANT   (b) Abilene

Fig. 3. Entropy of the four features during the large file transfer as seen in each network

This is a nice example of how an anomaly can hide in the network traffic. This anomaly illustrates that because of the traffic mix it might be impossible to detect an anomaly in a given network, whatever the value of the detection threshold is. Detecting an event hidden behind a predominant traffic requires either a different representation of the traffic, or detection in multiple networks. Summarizing the traffic as ingress-egress traffic matrix might separate the predominant traffic from the anomalous traffic and make it possible to detect the anomalous traffic.

4.3   Impact of Anonymization

Abilene anonymizes the IP addresses by inserting zeros on the last eleven bits. As explained in the previous section, it is difficult to evaluate the impact of such anonymization on the detection process. We did not find any anomaly that disappeared in Abilene because of anonymization. However, we found many instances of the following phenomenon, i.e. where anonymization impacts how the anomaly is classified.

On November 16th at 05:00 GMT, we detected a port scan from a university connected to Abilene in Atlanta, to a sub-network connected to the Swedish router in
GEANT. The maximum rate observed in Abilene was about 1,000 sampled packets
every five minutes. The ingress link in Abilene is lightly loaded so even with this rate
this anomaly was visible in the entropy of the features (fig. 4). The entropy of the port
numbers increase as the distribution is spread. The distribution of the source IP is concentrated around the attacker's IP address as visible in the decreased entropy. But the
entropy of the destination IP decreases indicating a concentration. In fact all the vic-
tims' IP addresses belong to the same sub-network and the side effect of anonymization
is to make them look like a single IP address, creating an artificial concentration in the
destination IP distribution.

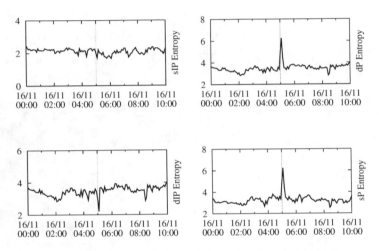

Fig. 4. Entropy of the four features during a Port Scan inside Abilene

This observation has multiple consequences. We can imagine non anomalous traffic
sent to multiple addresses in the same sub-network to be identified as an anomaly after
anonymization. That should be easy to detect though. On the other hand, we could
not imagine any scenario where anonymization could hide an anomaly. There is also a
good side-effect of anonymization. As mentioned in section 2.3, identifying the traffic
corresponding to an entropy increase (traffic dispersion) is usually very difficult. We
could imagine to use anonymization as a way to transform a traffic dispersion in a traffic
concentration. We keep the study of how to use anonymization techniques to help the
anomaly classification for future work.

5   Conclusion and Future Work

We have shown that numerous factors impact the detectability of traffic anomalies in a
given network. The major factors are detection methodology, data formalism, sampling
and network traffic. IP address anonymization on the other hand might end being a

feature that could make it easier to classify anomalies. However, its impact on anomaly detectability is still unclear. This work does not explain how each factor impacts the number and type of anomalies detected. However, it makes a clear case for (1) deeper analysis of anomaly detection techniques design and calibration and (2) Internet-Wide anomaly detection as a single method will not be capable to detect all anomalies in a network.

Using our two data sets, we are now starting a systematic analysis of two methods, Kalman and PCA, with different data formalisms, in order to understand how robust these techniques are and how to automatically choose the right operating parameters. Another important piece of work is to understand what is the minimum sampling rate in order not to miss anomalies. 1 for 1000 seems to be below that threshold.

A major concern is the lack of ground truth. We have started the annotation of the three weeks of traces used in this work. The annotated data set, including the anomalies we have detected, will be made available to the research community for comparison of observation and to facilitate the reproducibility of detection result, and the design of new detection techniques.

References

1. Lakhina, A., Crovella, M., Diot., C.: Diagnosing network-wide traffic anomalies. In: ACM Sigcomm. (2004)
2. Soule, A., Salamatian, K., Taft, N., Nucci, A.: Traffic matrix tracking using kalman filters. ACM LSNI Workshop (2005)
3. Soule, A., Salamatian, K., Taft, N.: Combining filtering and statistical methods for anomaly detection. In: ACM IMC. (Oct. 2005)
4. Zhang, Y., Ge, Z., Greenberg, A., Roughan, M.: Network anomography. In: ACM IMC. (Oct. 2005)
5. Teixeira, R., Duffield, N.G., Rexford, J., Roughan, M.: Traffic matrix reloaded: Impact of routing changes. In: PAM. (2005)
6. Brauckhoff, D., Tellenbach, B., Wagner, A., Lakhina, A., May, M.: The effect of packet sampling on anomaly detection. In: ACM IMC. (Oct. 2006)
7. Barakat, C., Iannaccone, G., Diot, C.: Ranking flows from sampled traffic. In: ACM CoNEXT. (Dec. 2005)

Leveraging BitTorrent for End Host Measurements

Tomas Isdal, Michael Piatek, Arvind Krishnamurthy, and Thomas Anderson

Department of Computer Science and Engineering
University of Washington, Seattle, WA, 98195

Abstract. Traditional methods of conducting measurements to end hosts require sending unexpected packets to measurement targets. Although existing techniques can ascertain end host characteristics accurately, their use in large-scale measurement studies is hindered by the fact that unexpected traffic can trigger alarms in common intrusion detection systems, often resulting in complaints from administrators. We describe *BitProbes*, a measurement system that works around this challenge. By coordinated participation in the popular peer-to-peer BitTorrent system, BitProbes is able to unobtrusively measure bandwidth capacity, latency, and topology information for ~500,000 end hosts per week from only eight vantage points at the University of Washington. To date, our measurements have not generated a single complaint in spite of their wide coverage.

1 Introduction

Detailed estimates of Internet path properties allow applications to improve performance. Content distribution networks (CDNs) such as Coral [1], CoDeeN [2] and Akamai [3] use topology information to redirect clients to mirrors that provide the best performance. Peer-to-peer services such as Skype [4] and BitTorrent [5] can use knowledge of both the core and the edge to optimize peer selection, improving end-user performance [6]. Overlay services such as RON [7] optimize routes based on metrics such as loss rate, latency, or bandwidth capacity to allow applications to select routes based on specific needs.

Although valuable, large-scale measurement of edge characteristics is encumbered by uncooperative or even hostile hosts. End hosts are often firewalled, making them unresponsive to active probing and ruling out many of the tools commonly used for mapping the Internet core. Further, administrators frequently mistake measurement probes for intrusion attempts, raising alarms in intrusion detection systems (IDSs). To avoid these problems, researchers have been forced to take alternate approaches when measuring end hosts. Systems such as PlanetSeer [8] take an *opportunistic* approach to end host measurement. PlanetSeer infers link properties by passively monitoring existing TCP connections between a centralized content provider and end hosts. By piggybacking measurements on existing, expected traffic, an opportunistic approach allows measurement of end hosts without triggering alarms in IDSs. Although successful, these systems are limited by the popularity of the service offered and require content providers to instrument servers with custom measurement software. This method has the additional drawback that the majority of data is transferred from the server to the end host, making accurate measurements of the end hosts upload capacity impossible.

S. Uhlig, K. Papagiannaki, and O. Bonaventure (Eds.): PAM 2007, LNCS 4427, pp. 32–41, 2007.

Drawing inspiration from these, we consider an alternative platform for attracting measurement targets: the BitTorrent peer-to-peer system. BitTorrent has several desirable features for such a platform. It is extremely popular, suggesting wide coverage. Also, its normal operation involves sending a large number of data packets over TCP connections in both directions, making those connections amenable to inspection by measurement tools.

We develop a tool aimed at leveraging BitTorrent for large-scale Internet measurements and make the following contributions:

- We present the design, implementation, and evaluation of *BitProbes*, a system that performs large scale measurements to end hosts using BitTorrent.

- BitProbes discovers ∼500,000 unique measurement targets in a single week, an order of magnitude more than previous systems.

- By using features of the BitTorrent protocol, BitProbes attracts traffic from end hosts that it uses to measure upload capacity, latency, and network topology. To the best of our knowledge, it is the first system that elicits TCP streams from non-webservers.

- We determine the bandwidth distribution for end hosts, updating a previous study of Gnutella users collected in 2002 [9].

- BitProbes collects comprehensive logs of BitTorrent protocol traffic and client behavior. We make anonymized versions of these logs public and present a surprising immediate fallout from their analysis in Sect. 4.5. Specifically, peer capacity is an uncertain predictor of performance in BitTorrent.

2 BitTorrent Overview

BitTorrent is a peer-to-peer file distribution tool that has seen a surge in popularity in recent years [10]. In this section, we describe the features of the BitTorrent protocol that make it amenable to end host measurements. A complete description of the BitTorrent protocol is beyond the scope of this paper, but is readily available [5,11].

For the purpose of Internet measurement, two parts of the protocol specification are relevant: how peers discover one another and exchange control information and when peers are permitted to send and receive file data.

2.1 Data and Availability Messages

BitTorrent is a request driven protocol. To initially connect to a swarm, a client first contacts a centralized coordinator called the *tracker*. The tracker maintains a list of currently active peers and provides a random subset to clients upon request. The client will then initiate TCP connections with the peers returned by the tracker. Clients then request small pieces (typically 64–512 KB) of the complete file from directly connected peers that possess them. When the download of any one small piece completes, a client is required to notify its connected peers that it has new data available for relay via a have message. By serving these small pieces of the whole file, peers assist in its distribution before completing the download themselves.

In addition to per-piece updates between directly connected peers, availability information is also exchanged after the handshake between two newly connected peers via a

`BitField` message. This message allows newly connected peers to efficiently update each other's views of pieces available for request.

By monitoring the `BitField` and `have` messages a peers broadcasts, it is possible to infer the download rate of that peer by multiplying the rate at which `have` messages are sent by the known size of each piece. We use this technique in the instrumented BitTorrent client of BitProbes, enabling measurement of not only the properties of end hosts but also information regarding BitTorrent application level behavior.

2.2   Choking, Unchoking, and Optimistic Unchokes

A noted feature of BitTorrent is its support for robust incentives for contribution, achieved via a tit-for-tat (TFT) reciprocation policy. Informally, a peer preferentially sends data to those peers that have recently sent it data. This has the effect that the more a peer contributes to the swarm, the faster the download rate of that peer will be. Before a peer Q is permitted to request data from a peer P, Q must receive an `unchoked` message from P, meaning that its requests are permitted and will be serviced. If, later, P notices that Q is sending data more slowly than other peers, P will send a `choke` message to Q, indicating that it can no longer make requests.

The problem with a pure TFT approach is that it provides no reason for an existing client to unchoke a newly joined peer as that peer does not have any pieces with which to reciprocate. To work around this difficulty and bootstrap new users, the BitTorrent protocol includes so-called *optimistic unchokes*. Every 30 seconds, each client randomly selects a member of its peer set to unchoke. This serves two purposes; it helps to bootstrap new peers so that they can contribute their resources. Also, it allows each peer to search the set of available peers for those willing to reciprocate with greater capacity. For the purpose of end-host measurements, optimistic unchokes are crucial.

3   BitProbes Design and Implementation

BitProbes provides a platform for conducting large-scale measurements of end hosts. By layering a measurement infrastructure on top of BitTorrent swarms associated with popular files, BitProbes leverages the willing participation of end-users to quickly discover measurement target. This section discusses the BitProbes architecture and the challenges associated with layering a measurement infrastructure on BitTorrent.

3.1   Attracting Traffic

The key to achieving broad coverage of end hosts is making sure that the BitTorrent swarms that BitProbes targets have a large number of participants. To this end, we crawl several popular websites that aggregate swarm connection information and user statistics. We rank these by total users and assign the most popular to measurement nodes running our instrumented BitTorrent client. We do not store or serve any content obtained. This allows BitProbes to connect to the full range of popular swarms since the risk of distributing copyrighted material is eliminated.

BitProbes relies exclusively on optimistic unchokes to induce measurement targets to willingly send large packets filled with piece data. As a result, the behavior of the BitProbes client is optimized to maximize optimistic unchokes. First, we increase the size of the directly connected peer set. Most BitTorrent implementations maintain 50–100 simultaneous connections per swarm. To increase the likely number of optimistic unchokes, BitProbes increases this limit to 1000. The actual connection count is limited by the number of users in each torrent, but a few hundred directly connected peers per torrent is not uncommon. Second, we connect to many peers briefly. During our measurements, we observed that many BitTorrent client implementations quickly unchoke new connections for a single round without reciprocation, but rarely thereafter. Fortunately BitProbes does not require a large amount of data to be able to make measurements. To limit the resources consumed, both in the swarm and for the nodes running BitProbes, we will disconnect a peer after receiving 2 MB of data.

A challenge is that the trackers from which BitProbes receives candidate peers often specify a minimum time between requests. This time is usually 10 minutes. To increase the number of peers to which each measurements node can connect, BitProbes maintains a *shadow tracker* to share peer information among measurement vantage points. Each measurement node relays all peer information obtained from trackers to the single shadow tracker. The shadow tracker is queried every minute by all measurement nodes to increase their lists of candidate targets. In addition to increasing the number of candidate targets for each measurement node, the shadow tracker also minimizes probing by preferentially selecting only those targets yet to be measured.

3.2   Performing Measurements

BitProbes is designed to provide a wide ranging set of measurement targets to any tool that operates on timed packets from TCP connections. Each measurement node runs a packet logging tool that records packets from BitTorrent TCP connections and passes them to a given measurement tool. Many of the available techniques for transparently measuring properties of Internet paths require an open TCP connection between the target and the measuring node.

To evaluate the feasibility of our approach, the current BitProbes implementation focuses on upload capacity estimation using the MultiQ tool. MultiQ is known to be reasonably accurate, providing 85% of measurements within 10% of actual capacity [12]. This accuracy comes at a price, however, as MultiQ requires traces from a "significant flow" to provide a prediction. Fortunately, the large data packets obtained from BitTorrent connections are sufficient. We use the `libpcap` library to record kernel-level timestamps of incoming packets for accuracy. The arrival times for MTU-sized packets are then supplied to the MultiQ application, which reports the estimated upload capacity. MultiQ is sometimes able to discover the capacity of multiple bottlenecks between the measurement node and the end-host. We assume that the lowest capacity measured is the access link capacity of the end host, even though it also can be due to a bottleneck within the network or the capacity of the measurement node. It is therefore important that the measurement nodes have a high capacity connection. To measure per-hop link latencies and route information, we are currently experimenting with a technique similar to that used by TCP-Sidecar [13]. By injecting extra packets with the IP record route option and varying TTL into an existing TCP stream, the probe packets appear as legitimate TCP packets both to the end host being probed and to any intermediate firewalls.

Although our extension of BitProbes to include this technique is ongoing, it serves as an example of another measurement technique that can be integrated into the BitProbes framework.

3.3   Analysis of BitTorrent Protocol Messages

In addition to end host behavior, layering our measurement infrastructure on BitTorrent allows us to collect data regarding the BitTorrent protocol itself and its behavior in the wild. For instance, we record trace data of `have` and `BitField` messages, allowing us to infer download rates of peers. Correlating this data with capacity measurements provides insight into the effectiveness of BitTorrent's TFT incentive mechanism. We discuss this in Section 4.5.

The trace logs are aggregated at a centralized database for easy access and a global view of swarm behavior. In addition to capacity measurements, we also log all protocol messages as well as tracker responses, peer arrivals and departures, and crawled swarm file sizes and popularity. We make these anonymized versions of logs public.

4   Results

This section discusses initial results obtained with the BitProbes prototype. We show that BitProbes quickly discovers a wide ranging set of measurement targets and that BitTorrent's opportunistic unchokes can furnish MultiQ with enough large packets to provide capacity measurements.

Our measurement testbed includes eight machines running at the University of Washington. Each of these were running 40 instances of the modified BitTorrent client. The results presented were collected between September 2nd and September 9th, 2006. Torrent aggregation sites[1] were crawled once every 12 hours to obtain candidate swarms.

4.1   Rate of Connections

BitProbes provides measurement targets in two forms. First, it collects a list of valid candidate target IP addresses and ports on which those IPs expect to receive TCP traffic. Then, if an IP has no firewall blocking, or if port forwarding is properly set up, we directly connect to the peer. Otherwise, in the case of a peer P behind a NAT, we are able to conduct measurements to P only when P receives our BitProbes client IP from the tracker and decides to connect to it. In practice, this occurs frequently and is therefore a significant benefit to BitProbes, since many end hosts use NATs. In this section, we consider the rate at which new candidate IPs are discovered, independently of whether those connections resulted in successful measurements or not. The rate at which new candidates are discovered is an optimistic measure of BitProbes coverage. First, it gives an upper bound on the number of possible measurements BitProbes can provide when running on our current testbed. Second, it indicates how quickly Bit-Probes exhausts the pool of available swarms and candidate peers made available from the swarm aggregation sites crawled.

---

[1] The websites used were http://thepiratebay.org and http://www.mininova.org

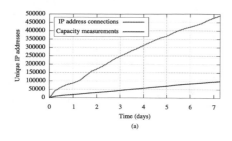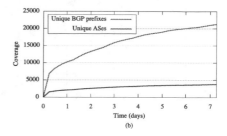

Fig. 1. (a): Number of connections and capacity measurements to unique IP addresses. (b): Increasing AS and BGP prefix coverage.

Figure 1(a) gives the number of distinct candidate IPs observed over the duration of our trace. During our week-long trace, the measurement nodes connected to or received connections from roughly 500,000 unique IP addresses. Because the rate of new connections is roughly constant, we conclude that the pool of end hosts participating in swarms listed on our crawled sites is significantly larger than the number that can be covered by our prototype system in a week. These results are promising given our plans for a wider deployment. By adding more measurement nodes to BitProbes, we expect to increase the number and coverage of measurements significantly.

As we log data for longer time periods, we expect an eventual decrease in the rate at which new end hosts are discovered. However, we are only crawling only two BitTorrent aggregation sites, and we have already discovered more popular BitTorrent swarm aggregation sites containing an order of magnitude more available swarms than sites we use currently. Further, crawling such sites is straightforward since RSS [14] feeds of available swarms are often provided.

4.2   Rate of Capacity Measurements

The raw number of connections provides an optimistic upper bound on the number of useful measurements possible with BitProbes. Providing concrete results, however, requires integrating an existing measurement tool into the BitProbes framework. We use the MultiQ tool for capacity measurements [12].

For MultiQ to infer the upload capacity of an end host, a TCP flow with a significant number of packets has to be sent by the end host to the measurement node. The packets are then analyzed by MultiQ in an attempt to provide a capacity measurement. Notably, MultiQ will fail rather than return a potentially erroneous measurement if the trace does not meet its correctness criteria. We provide MultiQ with all packet traces having more than 100 MTU sized packets. During our trace, significant flows from 176,487 unique IP addresses were received. In 96,080 of these cases (54%), MultiQ reported an upload capacity estimate.

As shown in Fig. 1(a), the number of measurements increases steadily throughout our measurement period, providing upload capacity measurements of roughly 100,000 unique IPs in total, meaning that ~20% of discovered end hosts provide targets amenable to measurement. In contrast to the upper bound provided by the total number of connections, these results provide a more conservative estimate of the likely number of targets that BitProbes can provide to measurement tools. To measure capacity, MultiQ requires specific sanity checks on the number and sizing of packets, requirements that are more stringent than what other tools might need. Further, obtaining TCP packet traces requires that end hosts peer with and optimistically unchoke BitProbes clients, while other techniques (such as TCP-Sidecar) might not require a packet trace.

4.3   Network Coverage

Having confirmed that BitProbes is capable of connecting to and collecting measurements from many thousands of distinct targets, we next turn to whether those targets cover the administrative and prefix diversity of the Internet. We chose to build on the popularity of BitTorrent because we expected its popularity to be reflected in wide coverage. In this section, we examine whether we achieve this.

During our week-long measurement period, BitProbes discovered candidate hosts in 21,032 unique BGP prefixes. A BGP prefix corresponds to an entry in the global BGP routing table. As seen in Fig. 1(b), the rate of new prefix observations diminishes over the course of the trace. Almost half of the total number of prefixes observed were discovered during the first 24-hour interval. Even though the swarms joined have a large number of end hosts associated with them, many of these hosts are in a limited subsection of the Internet at large. We suspect that this bias is due primarily to our limited selection of torrent aggregation sites, which cater primarily to English speaking audiences in the US and Europe.

Coverage of BGP Autonomous Systems (ASes) exhibits properties similar to those of BGP prefixes. Connections to a total of 3,763 ASes were observed, with roughly half being discovered in the first 12 hours of operation. Because AS coverage and BGP prefix coverage are correlated, we expect our efforts at increasing prefix coverage by diversifying our crawling to also increase AS coverage. Since one AS often corresponds to one Internet Service Provider (ISP) or POP, higher coverage again requires crawling swarms of global or diverse regional interest. However, the activity and size of ASes varies significantly. The bias in our coverage at the AS level reflects skew in size. Because we do not preferentially probe targets based on AS, we will simply have more measurements to popular ASes and fewer to those less popular. Nevertheless, our current sources of candidate swarms are sufficient to cover ~20% of total ASes in one week[2].

4.4   Capacity Distribution

Knowing the capacity distribution for end hosts on the Internet is a crucial building in the designing distributed systems. This section reports on the observed upload capacity distribution recorded during our evaluation of BitProbes, given in Fig. 2(a). Since the measurements are of hosts running BitTorrent, we point out that these results are not completely general. For instance, users with dial-up connectivity are unlikely to participate in the distribution of multi-gigabyte files via BitTorrent.

---

[2] We estimate the total number of ASes by counting unique observed AS numbers from merged RouteViews [15] and RIPE [16] traces.

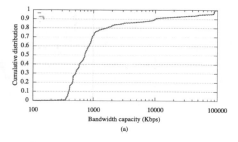
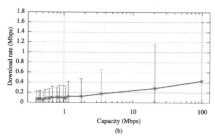

Fig. 2. (a): Upload capacity distribution of BitTorrent users. (b): The average download through-put received by our modified client from peers in a given capacity range. Although average throughput is correlated with upload capacity, no conclusive throughput prediction can be made.

A majority (70%) of hosts have an upload capacity between 350 Kbps and 1 Mbps. Only 10% of hosts have an upload capacity of 10 Mbps or more. However, the 5% of hosts with bandwidth capacities between 55 Mbps and 110 Mbps contribute 64% of the available resources, suggesting that successfully incorporating the resources of the high capacity clients is crucial for achieving high utilization in peer-to-peer systems. We make this capacity distribution public along with our BitTorrent trace data.

4.5   Capacity as a Performance Predictor in BitTorrent

This section discusses the applicability of our quick capacity estimation framework to BitTorrent itself. As discussed in Sect. 2.2, BitTorrent employs a TFT reciprocation strategy to encourage peers to contribute resources to the swarm. Although TFT tends to correlate contribution and performance, studies have suggested that in practice it often results in unfair peerings for high capacity users [17]. One suggested method for avoiding this unfairness has been to use quick bandwidth estimation techniques as a basis for selecting peering relationships in BitTorrent. In this section, we evaluate the ability of MultiQ's capacity estimates to predict observed upload rates from those peers from which we receive data.

Figure 2(b) gives the average of observed upload rates at which we receive piece data from peers as a function of the measured upload capacity of peers in a given capacity range. Error bars show the 5th and 95th percentiles of observed download rates of our modified client.

Because BitTorrent shares available upload capacity among several TCP connections simultaneously, we do not expect observed download rates as seen by a single peer to match capacity directly. However, for quick bandwidth estimation to select peers with good performance, the rank ordering of estimated bandwidth and observed performance should be correlated. This is not the case in BitTorrent today, with observed download rates varying quite a bit for for a given capacity (shown in Fig. 2(b)). From the perspective of a single peer in a single swarm, it is impossible to determine as to how many TCP connections a particular client has, whether the client's user has imposed application-level bandwidth limits, or whether the user is participating in multiple swarms simultaneously, all of which will cause actual transfer rate to be much lower than raw capacity.

Still, average performance tends to correlate with estimated capacity, suggesting that although quick bandwidth estimation cannot serve as a replacement for ranking peers by observed performance, it may provide early guidance for selecting peers in the absence of other information.

5 Related Work

The opportunistic measurement strategy of BitProbes is broadly similar to that of PlanetSeer [8], which monitor users of the CoDeeN CDN [2] to detect failures on the Internet. By monitoring existing TCP connections to CoDeeN users, it detects when a large number of clients simultaneously disconnect. After such events, PlanetSeer probes the remote nodes from different vantage points. If an endpoint is reachable from some vantage points but not others, it is recorded as a route abnormality. Using CoDeeN as a data source, PlanetSeer observes between 9,000–12,000 clients per day. Sherwood and Spring [13] note that when monitoring the CoDeeN network for a week, 22,428 unique IP addresses were seen using hundreds of available PlanetLab nodes as measurement hosts. In contrast, BitProbes observes 500,000 connections to unique IP addresses in a week from only eight vantage points.

The TCP-Sidecar project shares our goal of avoiding complaints from system administrators [13]. While BitProbes focuses on providing candidate targets to generic measurement tools, TCP-Sidecar is designed to construct an accurate router-level topology of the Internet using a modified traceroute tool. Since traceroutes can cause IDS alarms, modified traceroute packets are embedded into existing TCP connections. TCP-Sidecar use TCP connections from two sources: 1) passive monitoring of CoDeeN, resulting in measurements to 22,428 unique end-hosts per week and 2) downloading `robots.txt` files from web-servers, providing 166,745 unique IP addresses. Both of these sources of candidate targets are limited, in contrast, including TCP-Sidecar's topology measurement techniques in the BitProbes framework may dramatically increase their coverage and effectiveness.

Casado et. al. [18] examine unconventional sources of traffic including spam, worms, and automated scans, yielding a potential coverage of several hundred thousand IP addresses. The authors point out that CAIDA received probes from 359,000 infected servers during the first Code Red outbreak. Using tools similar to those of BitProbes, their system can infer link and path properties to the traffic sources. The number of significant flows from unique sources was limited, however, with only 2,269 unique sources during a 24 hour period. In comparison, BitProbes attracted 176,487 flows of more than 100 MTU size packets from unique sources during one week of operation. On the whole, BitTorrent appears to be a more fruitful source of candidate targets than other sources previously considered.

6 Conclusion

This paper describes BitProbes, a measurement system for end hosts on the Internet. BitProbes attracts connections from close to half a million unique end-hosts per week by leveraging the popularity of BitTorrent peer-to-peer filesharing swarms. The observed connections are used to infer link latency, topology, and capacity. All measurements are performed unobtrusively, avoiding probes that might trigger IDS alarms. Prototype results demonstrate that BitProbes provides close to an order of magnitude more candidate connections than previous systems relying on opportunistic measurement, suggesting the leverage of attracting measurement candidates via BitTorrent.

References

1. Freedman, M.J., Freudenthal, E., Mazières, D.: Democratizing content publication with Coral. In: NSDI. (2004)
2. Wang, L., Park, K., Pang, R., Pai, V.S., Peterson, L.L.: Reliability and security in the CoDeeN content distribution network. In: USENIX. (2004)
3. Akamai Inc: Akamai: The trusted choice for online buisniess. http://www.akamai.com (2006)
4. Skype: The whole world can talk for free. http://www.skype.com (2006)
5. Cohen, B.: Incentives build robustness in BitTorrent. In: P2PEcon. (2003)
6. Madhyastha, H.V., Isdal, T., Piatek, M., Dixon, C., Anderson, T., Krishnamurthy, A., Venkataramani, A.: iPlane: An information plane for distributed services. In: OSDI. (2006)
7. Anderson, D.G., Balakrishnan, H., Kaawhoek, M.F., Morris, R.: Resilient Overlay Networks. In: SOSP. (2001)
8. Zhao, M., Zhang, C., Pai, V., Peterson, L., Wang, R.: PlanetSeer: Internet path failure monitoring and characterization in wide-area services. In: OSDI. (2004)
9. Saroiu, S., Gummadi, P., Gribble, S.: Sprobe: A fast technique for measuring bottleneck bandwidth in uncooperative environments (2002)
10. Parker, A.: The true picture of p2p filesharing. http://www.cachelogic.com/home/pages/studies/2004_01.php (2004)
11. Cohen, B.: BitTorrent Protocol Specifications v1.0. http://www.bittorrent.org/protocol.html (2002)
12. Katti, S., Katabi, D., Blake, C., Kohler, E., Strauss, J.: MultiQ: Automated detection of multiple bottleneck capacities along a path. In: IMC. (2004)
13. Sherwood, R., Spring, N.: Touring the Internet in a TCP Sidecar. In: IMC. (2006)
14. RSS Advisory Board: Really Simple Syndication: RSS 2.0 Specification. http://www.rssboard.org/rss-specification (2006)
15. Meyer, D.: RouteViews. http://www.routeviews.org (2005)
16. RIPE NCC: Routing Information Service. http://www.ripe.net/ris/ (2006)
17. Bharambe, A., Herley, C., Padmanabhan, V.: Analyzing and Improving a BitTorrent Network's Performance Mechanisms. In: IEEE INFOCOM. (2006)
18. Casado, M., Garfinkel, T., Cui, W., Paxson, V., Savage, S.: Opportunistic measurement: Extracting insight from spurious traffic. In: HotNets. (2005)

Trace Driven Analysis of the Long Term Evolution of Gnutella Peer-to-Peer Traffic*

William Acosta and Surendar Chandra

University of Notre Dame, Notre Dame IN, 46556, USA
{wacosta,surendar}@cse.nd.edu

Abstract. Peer-to-Peer (P2P) applications, such as Gnutella, are evolving to address some of the observed performance issues. In this paper, we analyze Gnutella behavior in 2003, 2005, and 2006. During this time, the protocol evolved from v0.4 to v0.6 to address problems with overhead of overlay maintenance and query traffic bandwidth. The goal of this paper is to understand whether the newer protocols address the prior concerns. We observe that the new architecture alleviated the bandwidth consumption for low capacity peers while increasing the bandwidth consumption at high capacity peers. We measured a decrease in incoming query rate. However, highly connected *ultra-peers* must maintain many connections to which they forward all queries thereby increasing the outgoing query traffic. We also show that these changes have not significantly improved search performance. The effective success rate experienced at a forwarding peer has only increased from 3.5% to 6.9%. Over 90% of queries forwarded by a peer do not result in any query hits. With an average query size of over 100 bytes and 30 neighbors for an ultra-peer, this results in almost 1 GB of wasted bandwidth in a 24 hour session. We outline solution approaches to solve this problem and make P2P systems viable for a diverse range of applications.

1 Introduction

In recent years, Peer-to-Peer (P2P) systems have become popular platforms for distributed and decentralized applications. P2P applications such as Gnutella[1], Kazaa [5] and Overnet/eDonkey [6] are widely used and as such, their behavior and performance have been studied widely [11,3,8]. Gnutella, although popular, has been shown to suffer from problems such as free-riding users, high bandwidth utilization and poor search performance. With the number of users in the network growing and the ultra-peer architecture becoming more dominant [9], solving the problems of bandwidth utilization and search performance becomes increasingly important to the long-term viability of Gnutella as well as other filesharing P2P applications.

In this paper we present an analysis of the trends in Gnutella traffic 2003, 2005, and 2006. First, we examined large-scale macro behaviors to determine how

---

* This work was supported in part by the U.S. National Science Foundation (IIS-0515674 and CNS-0447671).

S. Uhlig, K. Papagiannaki, and O. Bonaventure (Eds.): PAM 2007, LNCS 4427, pp. 42–51, 2007.

Gnutella traffic evolves over the long term. We captured traffic in 2003, 2005, and 2006 and compare the message rate, bandwidth utilization, and queueing properties of each Gnutella message type to determine the changes in characteristic behavior of Gnutella traffic over the long term. We found that the percentage of the bandwidth consumed by each message type changed considerably. We also studied more localized behavior such as the query success rate experienced by peers forwarding queries into the network. We found that the overwhelming majority ($> 90\%$) of the queries forwarded by a peer do not yield any query hits back to the forwarding peer. We show that earlier changes to the Gnutella protocol had not achieved the intended benefits of alleviating poor search performance and high bandwidth utilization.

2   Gnutella Overview

First we describe the Gnutella system. Gnutella is a popular P2P filesharing application. Gnutella users connect to each other to form an overlay network. This overlay is used to search for content shared by other peers. The Gnutella protocol is a distributed and decentralized protocol. Peers issue request messages that are flooded to all of their neighbors. Each neighboring peer, in turn, forwards the request to all of its neighbors until a specified time-to-live (TTL) is reached. This flooding mechanism allows for reaching many peers, but can quickly overwhelm the available network bandwidth. Next we describe the two major versions of the Gnutella protocol.

2.1   Protocol v0.4

The original Gnutella protocol, v0.4, assumed that each Gnutella peer was equal in terms of capacity and participation. The v0.4 protocol specified four major messages: Ping, Pong, Query, and QueryHit. The protocol also specified other messages such as Push requests, but these messages were rare and did not represent a significant percentage of the messages sent in the Gnutella network. The Ping and Pong control messages are used by Gnutella clients to discover other peers in the network. Ping and Query messages are flooded to each neighbor. The flooding continues until the TTL for the request has been reached. Responses such as Pong and QueryHit messages are routed back to the originator along the path of the request. When a peer receives a Query message it evaluates the query string and determines if any of its shared content can satisfy the query. If so, the peer sends back a QueryHit message along the path the it received the Query. Each QueryHit message contains the following information: the number of objects that match the query, the IP address and port of the peer where the objects are located, and a set of results including the file name size of each matching object.

2.2   Protocol v0.6

As Gnutella became popular, both the size of the network and the amount of traffic on the network increased. The increase in network traffic coupled with the poor Internet connections of many users [11] overwhelmed many peers in terms of routing and processing messages. In order to overcome the performance limitations of the original protocol, the v0.6 [2] was introduced. The new v0.6 ultrapeer architecture reduced the incoming query message rate thus lowering the bandwidth requirements to process queries. However, as we show in this paper, the architecture change did not eliminate the problem of high bandwidth consumption; it shifted it to a different place in the network. Specifically, the v0.6 protocol requires the ultrapeers to maintain many connections.

3   Related Work

Saroiu et. al. [11] and Ripeanu et. al. [10] studied several aspects of the Gnutella file-sharing network. The authors discovered that new clients tended to connect to Gnutella peers with many connections. This led to an overlay network whose graph representation had a node degree distribution of a power law graph. The authors also identified that most users do not share content (free-ride) and that a large percentage of nodes had poor network connectivity. More recently, the problem of free-riding on Gnutella was revisited in [4]. The authors found that a greater number of users, as a percentage of the total users, are free-riding on Gnutella than in 2000. Queries forwarded to free-riding users will not yield any responses. Thus free-riding users degrade the high-level utility of the system and the low-level performance

A more closely related work to this paper is by Rasti et al. [9]. Their aim was to analyze the evolution of the v0.6 two-tier architecture with respect to the topological properties of the network. The authors measured an increase in the number of nodes in the network and showed that as nodes began to transition to the v0.6 protocol, the network became unbalanced with respect to the number of leaf and ultra-peers. They showed that modifications to major Gnutella client software and rapid adoption rate by users helped restore the overlay's desired properties. Our work is similar in that we analyze Gnutella's evolution from the v0.4 to the v0.6 architecture. However, our work focuses on the evolution of traffic characteristics and query performance. This allows for better understanding of the effects of architectural changes to the system. Further, understanding the evolution of the system's workload enables better decisions about future changes to the system and facilitates better modeling and simulation of large-scale P2P networks.

4   Results

4.1   Methodology

We captured traces for two weeks in May and June of 2003 as well as October of 2005, and June through September of 2006. We modified the source code of Phex [7], an open source Gnutella client written in Java to log traffic on the network. The client would connect to the network and log every incoming and outgoing message that passed through it. It is important to note that our traffic capturing client does not actively probe the system; it only logs messages that it receives and forwards. Each log entry contained the following fields: timestamp, Gnutella message ID, message direction (incoming/outgoing), message type, TTL, hops taken, and the size of the message. In addition to logging incoming and outgoing messages, the client also logged queries and query responses in separate log files. For queries, we logged the query string. In the case of query responses, the log contained the Gnutella message ID of the original query, the peer ID where the matching objects are located and a list of the matching objects which includes the filename and the size of the file. The original protocol did not have separate classes of peers. As such, the traces from 2003 were collected with our client running as a regular peer. With the introduction of the v0.6 protocol, peers were classified as either leaf or ultra-peers. Our traces from 2005 and 2006 were captured with our client acting as an ultra-peer. The client was allowed to run 24 hours a day and logs were partitioned into two hour interval for ease of processing. These traces give insight as to the evolution of the Gnutella traffic over the last several years. We use the data from these traces to evaluate our design choices. In the interest of space, we present only results from one 24 hour trace from a typical weekday. The same day of the week was used (Thursday) for the traces of each year. Analysis of other days show similar results as those presented in this paper with weekends showing only 3% more traffic than weekdays.

4.2   Summary of Data

The average number of messages (incoming and outgoing) handled by the client was 2.5M in 2003 and 2.67M in 2006 per 2 hour interval. The average file size for a 2 hour trace was 292MB in 2003 and 253MB in 2006. This represents over 3GB of data per 24 hour period. In 2003, a 24 hour trace saw 2,784 different peers directly connected to it. In 2006, our client only saw 1,155 different peers directly connected to it in a 24 period. Although our 2006 traces saw fewer peers, other studies [8,9] show that the network has grown over time. The reduction in number of peers seen by our client can be attributed to a trend toward longer session times. In 2003, more than 95% of the sessions lasted less than 30 minutes. In 2006, more than 50% of the sessions lasted longer than 60 minutes. Longer session times imply slower churn rate for the connections at each node and thus our client would not see as many connections when the session times for peers is longer.

It is important to note that our data from 2005 shows characteristics that are consistent with the evolution of the protocol with respect to TTL and hops taken for messages. However, the bandwidth data is somewhat skewed. Investigation of this data revealed that although the client attempted to act as an ultra-peer, it was not able to maintain many connections (< 10). In contrast, the same client in 2006 was able to maintain over 30 connections consistently. Our traces from a client running as a leaf node show different behavior with a significant reduction in bandwidth for leaf nodes in 2006.

Table 1. Mean TTL left and Hops Taken for different message types in a 24 hour period from 2003, 2005, and 2006 data

	All Messages		Control Messages		Query Messages	
	TTL Left	Hops Taken	TTL Left	Hops Taken	TTL Left	Hops Taken
2003	2.39	4.46	2.84	3.11	2.36	4.57
2005	2.95	3.41	3.17	3.76	2.52	2.71
2006	0.52	3.34	0.37	3.42	2.27	2.47

4.3   Message TTL Analysis

The TTL and the number of hops taken for each give an indication of how far messages travel in the network. Table 1 shows the mean TTL and number of hops taken for different classes of messages in a 24 hour trace for 2003, 2005 and 2006 traffic. In 2003, messages travel 6.85 hops into the network. Messages required 4.46 hops to reach our logging client and arrived with a mean TTL left of 2.39. Similarly, in 2005, messages travel 6.31 hops into the network requiring 3.41 hops to reach our client and arriving with a TTL of 2.95. In contrast, traffic from 2006 travels 3.86 hops into the network requiring 3.34 hops to reach our client and arriving with a mean TTL left of 0.52. This change can be attributed to the v0.6 protocol. The new protocol typically restricts the initial TTL of the message to between 4 and 5 hops to reduce the burden on the network due flooding from ultra-peers with many neighbors. A closer inspection of the TTL left and hops taken for the different types of messages reveals a change in the traffic characteristics of the different message types. In 2003, control messages (Ping and Pong) traveled further into the network and had to be propagated to many peers. On average, 2003 control traffic had a mean of 3.1 hops taken and a mean TTL left of 2.8. In contrast, 2006 control traffic had a mean of 3.4 hops taken and a mean TTL left of 0.37. Although messages arrive at a node from similar distances in 2003 and 2006, they are not expected to be propagated as far in 2006 as they were in 2003. Unlike control traffic, query traffic in 2006 is expected to be propagated further at each node. In addition, the TTL left for query traffic has remained fairly stable from 2003 to 2006: 2.36 in 2003 and 2.26 in 2006.

We should note the trend between 2003, 2005 and 2006. In 2005, control messages traveled deep into the network (6.93 hops) compared to 2006 (3.79 hops). In contrast, query messages traveled 6.93 hops in 2003, 5.23 hops in 2005 and 4.68 in 2006. This shows that query messages did not travel as deep into the network in 2005 as they did in 2003. This can be attributed to the transitioning to the current state of the v0.6 protocol. The protocol specification only sets guidelines as to what values to set for the initial TTL of a message. In 2005, vendors of Gnutella clients had already begun to set lower TTL values for query messages, but had continued to use a TTL of 7 for control messages. In 2006, these vendors transitioned to lower TTL values for all messages. In Figures 1 and 2 we show the hourly mean TTL and hops taken for control and query traffic respectively. The traffic represents data from 24 hour traces from 2003 and 2006. We note that the TTL left and hops taken for each message remain relatively constant throughout the day even though bandwidth consumption and the number of nodes in the network fluctuate throughout the day. This observation can be used to develop traffic models when simulating a P2P file-sharing workload.

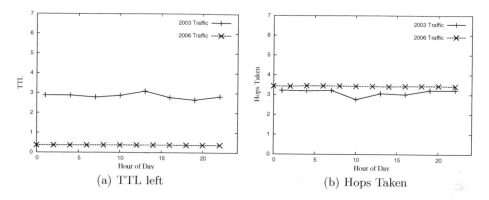

(a) TTL left   (b) Hops Taken

Fig. 1. Mean hourly TTL and hops taken for incoming control messages in a 24 hour period for 2003 and 2006 traffic

4.4   Bandwidth Analysis

In this section we analyze the traces to identify changes in bandwidth consumption. We track the bandwidth utilization for control (Ping and Pong), query (queries and query hits), and route table update messages. We limit the control messages only to the basic Ping and Pong messages as they make up the overwhelming majority of non-query messages. Additionally, by tracking route table updates separate from the basic control messages, we can determine how the behavior of the network has evolved with respect to the use of each message type. Note that we do not show bandwidth consumed by peers for downloading objects.

We examined bandwidth utilization by tracking the bandwidth requirements for each message type. Figure 3 shows the bandwidth utilization for each message type in a 24 hour period for 2003 and 2006 traffic respectively. In 2003, query traffic dominates the bandwidth utilization. Further, the amount of query traffic varies at different times of the day while control traffic remains relatively stable throughout the entire day. The 2006 traffic exhibits different properties. First, the query traffic represents significantly less bandwidth than in 2003, although out going query traffic is still the dominant consumer of bandwidth. Control traffic, both incoming and outgoing, plays a larger role under the current v0.6 protocol in 2006 than it did in 2003, as does route table update traffic. The changes in bandwidth throughout the day are most pronounced in incoming and outgoing

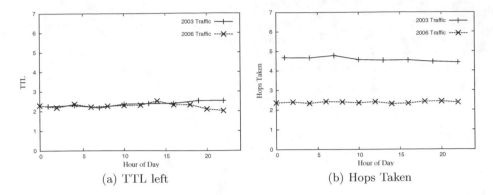

(a) TTL left   (b) Hops Taken

Fig. 2. Mean hourly TTL and hops taken for incoming query messages in a 24 hour period for 2003 and 2006 traffic

query traffic while control traffic remains relatively constant throughout the day. The increase in route table update traffic can be attributed to the fact that in 2003, very few clients were using the new v0.6 protocol. As more clients moved to the new protocol, we see a rise in the number of route table update messages sent in the system.

The evolution of the Gnutella protocol and network results in traffic characteristics that are different in 2006 than they were in 2003. First, in 2003, control and route table updates represent a small percentage of the total bandwidth utilization. In 2006, however, control traffic is significantly more prominent with respect to the percentage of bandwidth utilization. Additionally, route table updates become a non-trivial message category in 2006 whereas these messages were virtually non-existent in 2003. Finally, the characteristics of query traffic from 2003 to 2006 change dramatically. In 2003, query traffic (both incoming and outgoing) dominates the bandwidth utilization. In 2006, query traffic still consumes a large percentage of the bandwidth. However, only outgoing query traffic consumes a large amount of bandwidth. Incoming query traffic uses less bandwidth than control traffic in 2006. This is a result of the evolution of the protocol specification. The v0.6 protocol establishes ultra-peers as peers with many neighbors. Ultra-peers can collect information about files shared at their leaf nodes and use this information to respond to queries. We saw earlier that message TTL are decreased in the v0.6 protocol, so fewer messages are sent in the network. Additionally, the v0.6 query routing protocol attempts to minimize the number of query messages that are sent. Therefore, a node under the current v0.6 protocol will receive fewer incoming query messages than it would have in 2003. Because ultra-peers have many connections, propagating queries to each neighbor results in outgoing query traffic utilizing a large percentage of the total bandwidth. Although the outgoing query bandwidth is large in absolute terms, there is an improvement relative to 2003 since the outgoing query bandwidth

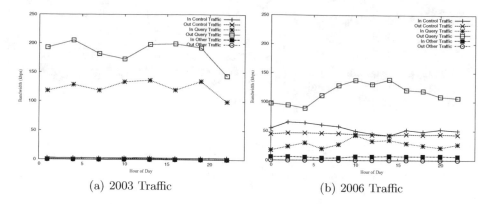

(a) 2003 Traffic   (b) 2006 Traffic

Fig. 3. Hourly bandwidth for different message types in a 24 hour period for 2003 and 2006 traffic. Note that the curves in Figure 3(a) do not wrap around to the same value at the end of a 24 hour period. This was due to our gathering procedure in 2003 that required the client to shut down after each 2 hour trace for processing. The client did not get restarted again for another hour in order to accommodate processing. This problem was fixed in our later data capturing process.

was reduced from 200 kbps to 100 kbps when the node is an ultra-peer with 30 connections. Leaf peers benefit from this situation as they have fewer connections and receive very few incoming queries.

4.5   Query Traffic Analysis

In the previous section we showed the evolution of bandwidth utilization for the different message types from 2003 to 2006. In this section, we will show the evolution of query traffic. We are interested in examining the relationship between the number of query messages sent and received and the success rate. Table 2 shows a summary of the query traffic measurements for traces from 2003 and 2006 traffic. The values in the table are for a typical 24 hour interval. In 2003, query traffic constituted a large component of the bandwidth used. We see in the table that in 2003, a peer would receive over 5M query messages in a 24 hour interval, or approximately 60 queries per second. In 2006, this number is significantly reduced to 280K queries in a 24 hour interval, or about 3 queries per second. With a mean query message size of 73.25 bytes for 2003 traffic, 60 queries per second corresponds to an incoming data rate of 4.39 KB/s or 35.2 kbps. Such a data rate would overwhelm most home dial-up connections. The current state of the network has reduced the number of queries received at each node at the expense of requiring a large number of connections for ultra-peer nodes. The large number of connections, in turn, results in a large outgoing bandwidth utilization. On a per query basis, 2003 traffic generated 0.285 KB per query, while 2006 traffic generated 4.026 KB per query. In 2003, each query was propagated to less than 4 peers on average while in 2006, each query was

Table 2. Query traffic summary for a 24 hour trace from 2003 , 2005 and 2006

	2003	2005	2006
Queries Received	5,261,064	614,365	279,235
Mean Queries per second	60.89	7.11	3.23
Query Message Size (including header)	73.25	69.04	105.61
Successful Queries	184,140	63,609	19,443
Success Rate	3.5%	10.3%	6.9%
Mean Queue Time (successful queries)	62.724 s	86.199s	5.435 s
Mean Outgoing Messages per Query	3.59	9.81	38.439
Mean Outgoing Bytes per Query	0.285 kB	0.677KB	4.026 kB
Mean Outgoing Query Bandwidth	138.82 kbps	38.5 Kbps	104.03 kbps

propagated to a mean of 38 peers. Note that in 2005, the bandwidth utilization is much lower compared to both 2003 and 2006. This is because our client was not able to operate as an ultra-peer and thus only was able to maintain less than 10 connections. However, other metrics such as incoming query rate for 2005 is consistent with the trend from 2003 to 2006.

Next we investigate the ability of the network to successfully resolve queries. We see that from 2003 to 2006, the success rate almost doubles from 3.5% in 2003 to 6.9% in 2006. Nevertheless, the success rate is still remarkably low. In 2006, a success rate of 6.9% implies that 93.1% of queries that reach a node will not be resolved after the node propagates the query. This means that each node is utilizing bandwidth to process queries but that effort is wasted on over 90% of those queries. In a 24 hour interval, a node receives approximately 280K queries with an average size of 105 bytes. Each of these queries is propagated to approximately 38 neighbors resulting in 1.14 GB of outgoing data, of which 93.1%, or 1.03 GB, are wasted since the success rate is only 6.9%.

5   Limitations

The study described in this paper was conducted using a single ultrapeer on our university's campus. As such, it may not be representative of the global behavior of Gnutella. We had performed measurements from a peer on a broadband network using a different service provider and observed similar results to as was reported in this paper. Also, the broadband studies were conducted in 2006; we do not have any measurement from 2003 or 2005 for this analysis. Hence, we cannot ascertain the consistency of the results among broadband users.

6   Conclusion

We presented the results of our measurement and analysis of the Gnutella file-sharing network. We showed that although the Gnutella architecture changed from 2003 to 2006 to help alleviate query bandwidth utilization, the success rate of queries has not shown significant improvements. Additionally, the bandwidth utilization problem is alleviated at low-capacity peers (leaf peers), but high capacity peers (ultra peers) experience an increase in query bandwidth utilization. The increase in bandwidth occurs due to the large number of neighbors connected to the ultra peers. These findings indicate that despite the efforts to improve the performance of Gnutella, search performance is limited by the design of the protocol. We are investigating an approach that exploits the underlying characteristics of the queries and the distribution of objects in the system in order to improve search performance.

References

1. Gnutella protocol v0.4. http://dss.clip2.com/GnutellaProtocol04.pdf.
2. Gnutella protocol v0.6. http://rfc-gnutella.sourceforge.net/src/rfc-0_6-draft.html.
3. Krishna P. Gummadi, Richard J. Dunn, Stefan Saroiu, Steven D. Gribble, Henry M. Levy, and John Zahorjan. Measurement, modeling, and analysis of a peer-to-peer file-sharing workload. In *Proceedings of the nineteenth ACM symposium on Operating systems principles*, pages 314–329. ACM Press, 2003.
4. Daniel Hughes, Geoff Coulson, and James Walkerdine. Free riding on gnutella revisited: The bell tolls? *IEEE Distributed Systems Online*, 6(6), June 2005.
5. Kazaa media desktop. http://www.kazaa.com/us/index.htm.
6. Overnet. http://www.overnet.org/.
7. The phex gnutella client. http://phex.kouk.de.
8. Yi Qiao and Fabin E. Bustamante. Structured and unstructured overlays under the microscope - a measurement-based view of two p2p systems that people use. In *Proceedings of the USENIX Annual Technical Conference*, 2006.
9. Amir H. Rasti, Daniel Stutzbach, and Reza Rejaie. On the long-term evolution of the two-tier gnutella overlay. In *IEEE Golbal Internet*, 2006.
10. M. Ripeanu, I. Foster, and A. Iamnitchi. Mapping the gnutella network: Properties of large-scale peer-to-peer systems and implications for system design. *IEEE Internet Computing Journal*, 6(1), 2002.
11. Stefan Saroiu, P. Krishna Gummadi, and Steven D. Gribble. A measurement study of peer-to-peer file sharing systems. In *Proceedings of Multimedia Computing and Networking 2002 (MMCN '02)*, San Jose, CA, USA, January 2002.

LiTGen, a Lightweight Traffic Generator: Application to P2P and Mail Wireless Traffic

Chloé Rolland[1], Julien Ridoux[2], and Bruno Baynat[1]

[1] Université Pierre et Marie Curie – Paris VI, LIP6/CNRS, UMR 7606, Paris, France
[2] ARC Special Research Center for Ultra-Broadband Information Networks (CUBIN),
an affiliated program of National ICT Australia
The University of Melbourne, Australia
{rolland,baynat}@rp.lip6.fr, j.ridoux@ee.unimelb.edu.au

Abstract. LiTGen is an easy to use and tune open-loop traffic generator that statistically models wireless traffic on a per user and application basis. We first show how to calibrate the underlying hierarchical model, from packet level capture originating in an ISP wireless network.[1] Using wavelet and semi-experiments analysis, we then prove LiTGen's ability to reproduce accurately the captured traffic burstiness and internal properties over a wide range of timescales. In addition the flexibility of LiTGen enables us to investigate the sensitivity of the traffic structure with respect to the possible distributions of the random variables involved in the model. Finally this study helps understanding the traffic scaling behaviors and their corresponding internal structure.

Keywords: traffic generator, scaling behaviors, energy plot, semi-experiments.

1 Introduction

The limited resources of wireless access networks, the users' contracts diversity and mobility are particularities that greatly impact the design of traffic models. Traffic generators proposed in the past years modeled primarily web traffic. [1] and [2] proposed hierarchical models, but did not validate them against real traffic traces. Recently, [3] is an effort to generate representative traffic for multiple and independent applications. The model underlying this generator is not designed to specify the packet level dynamics neither to capture the traffic scaling structure. In [4], the authors argue that network characteristics must be emulated to reproduce the burstiness observed in captured traffic. Their traffic generator relies then on a third party, link and network layers emulator (requiring the use of 11 cutting-edge computers). Thus, this opaque emulator makes the investigation of the obtained traffic scaling structure more complex.

---

[1] This study would not have been conducted without the support of Sprint Labs. The authors would like to thank Sprint Labs for giving access to the wireless traffic traces and particularly Ashwin Sridharan for his support.

S. Uhlig, K. Papagiannaki, and O. Bonaventure (Eds.): PAM 2007, LNCS 4427, pp. 52–62, 2007.

In this paper, we present LiTGen, a "**L**ight **T**raffic **Gen**erator" that statistically models wireless traffic. LiTGen relies on a simple hierarchical description of traffic entities, most of them modeled by uncorrelated random variables and renewal processes. The confrontation of LiTGen to real traces captured on an operational wireless network proves its ability to reproduce accurately, not only the observed traffic scaling behaviors over a wide range of timescales, but also the intrinsic properties of the traffic. This design does not require to consider network or protocol characteristics (*e.g.* RTT, link capacities, TCP dynamics...) and allows fast computation executed on a commonplace computer. To the best of our knowledge, we are the first ones to produce synthetic wireless traffic that accurately reflects the first two orders of the packets arrivals time series.

In the rest of this paper, section 2 describes LiTGen, its underlying model and how it generates synthetic traces. Section 3 validates LiTGen's ability to reproduce the complexity of the original traffic correlation structure, for both mail and peer-to-peer (P2P) traffics. We then investigate in section 4, the sensitivity of the traffic structure with respect to the distributions of the random variables involved in the underlying model. Finally we conclude this paper with a summary of our findings and directions for future work.

2 Building a Lightweight Generator

2.1 Underlying Model

Earlier works identified three possible causes of correlation in IP traffic: the presence of heavy-tailed distributions [5], the superimposition of independent traffic sources [6] and the inherent structure and interactions of protocol layers [7]. These two last assumptions call on the conception of our traffic generator to be based on a user-oriented approach and a hierarchical model. This model is made of several semantically meaningful levels, each of them characterized by a specific traffic entity. For each traffic entity, we define a set of random variables either related to a time or a size characterization.

Session level. We assume each user undergoes an infinite succession of session and inter-session periods. During a session, a user makes use of the network resources by downloading a certain number of objects. We define two random variables to characterize this level: $N_{session}$, the session size, *i.e.* the number of objects downloaded during a session and, T_{is}, the inter-session duration.

Object level. A session is made of one or several objects. Indeed, a session is split up into a set of requests (sent by user) and responses (from the server), where responses gather the session's objects. In the case of web, objects may be web pages' main bodies (HTML skeletons) or embedded pictures [8][2]. In the case of mail, objects may be servers responses to clients requests (*e.g.* e-mails, clients

---

[2] In this previous study applied to web traffic, the underlying model was made of four levels, including ***web pages level***. This extra level, not described here, is not relevant in the context of mail and P2P, but is kept for the generation of web traffic.

Fig. 1. Steps for filtering P2P traffic and identifying per-user traffic entities

accounts meta-data...). In the case of P2P, objects may be files or chunks of files. The description of this level requires the definition of two random variables: N_{obj}, the object size, *i.e.* the number of IP packets in an object and, IA_{obj}, the objects inter-arrival times in a session.

Packet level. Finally, each object is made of a set of packets. The arrival process of packets in an object can be described by giving the successive inter-arrival times between packets, characterized by random variables IA_{pkt}.

So far, we made no assumption concerning the random variables correlation structure. Indeed, inter-dependence mechanisms can be taken into account by introducing correlations between random variables. Of course, the objective here is to remain as simple as possible and to introduce dependencies only if necessary, as discussed in section 3. Note that one can equivalently remove from the hierarchy the session level by including the inter-session durations in the objects inter-arrivals distribution. Nevertheless, it would make the characterization of IA_{obj} more complex and LiTGen less easy to use in practice.

2.2   Wireless Trace Analysis

In order to calibrate and validate LiTGen underlying model, we benefit from data traces captured on the Sprint PCS CDMA-1xRTT access network. Traces have been captured on an OC-3 collecting link spanning a large geographical area and so tens of wireless access cells. The traffic capture consists in two unidirectional 24 hours long traces, captured simultaneously. Each of them is composed of a collection of IP packets with accurate timestamps and entire TCP/IP headers. Thus, we have access to the well-known 5-tuple: {IP destination, IP source, port destination, port source, transport protocol}. These traces have already been used in a previous study [9] that gives more details on the raw characteristics of this data traffic and its differences with wireline traffic.

Because of its small representation (less than 10%) in the traces and to narrow down the analysis, we exclude the UDP traffic from our study and focus on TCP traffic. Moreover, we are not interested in the modeling of the interactions between the upload and the download traffics. Indeed, we want to keep a very simple underlying model which do not rely on a network or TCP emulator. Finally, we focus on the traffic intended to the wireless terminals (download path). As a matter of fact, the upload wireless traffic contains mostly connection requests and ACKs, while the download wireless path is richer and has more importance from an operational point of view.

The model calibration requires to identify the characteristics of the traffic entities from the captured set of packets. To do so, we first filter traffic corresponding to a given application and then identify per-user traffic entities based on the 5-tuple associated to each packet (see Figure 1).

Packet Level. A filter based on a source port number selection retains traffic specific to a given application (*e.g.* 110, 143, 220 for the mail traffic). A user's packets share the same destination IP address and are then grouped into subsets of a given {IP source, port destination} pair, corresponding typically to the server IP address contacted and the port opened on the user's side. All resulting subsets of a given user correspond then to all flows he requested (considering the given application).

Object Level. After applying the application and user filters, packets subsets are grouped to identify objects by means of the method presented in [10]. Based on the analysis of the TCP headers, this method observes the TCP flags (SYN, FIN, etc) to differentiate objects within packets subsets (characterized by a rupture in the acknowledgment number series).

Session Level. Finally, we aggregate objects into sessions. The definition of the sessions relies on *active periods* during which one or several objects are being downloaded. Those periods are separated by *inactive periods*. We use a temporal clustering method (also used in [1,2,10] for the retrieving of web pages) to infer the sessions' boundaries. An inactive period that lasts for more than a predefined threshold determines the precedent session termination. We fixed empirically the threshold to 300 seconds[3].

2.3 Traffic Generation

Contrarily to the trace analysis, LiTGen generates traffic from upper level entities (sessions) to lower ones (packets). LiTGen is used for the generation of traffic corresponding to different user's applications. For each kind of traffic, we first fix the number of users of the corresponding application. LiTGen generates then traffic for each user independently. The final synthetic trace is obtained by superimposing synthetic traffic of all users and all applications. For validation purposes we can extract the proportion and the number of users of each application from the captured trace. In an operational network these statistics can be derived from operator's knowledge of customer's subscribes services[4].

3   Validation

LiTGen is evaluated on its ability to reproduce the complexity of the traffic correlation structure in the captured packet traces. For this purpose, we use an

---

[3] Note that the value of this threshold does not impact significantly the results.

[4] Such a finite assumption is typically used for network planning to predict the active population that will be served during a given time.

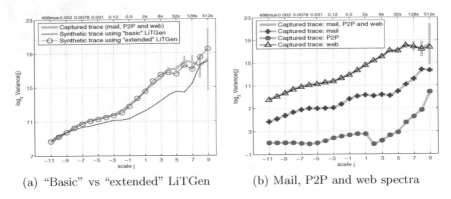

(a) "Basic" vs "extended" LiTGen   (b) Mail, P2P and web spectra

Fig. 2. Model evaluation and comparison of the mail, P2P and web spectra

energy spectrum comparison method to match the packets arrivals time series extracted from the original and corresponding synthetic traces. Since the 24-hour trace is not stationary, the analysis is performed on a one-hour period extracted from the entire trace. The results presented here correspond to a given one-hour period; similar results were obtained for other one-hour extracted traces.

3.1   Wavelet Analysis

We use the Logscale Diagram Estimate or LDE [11] to perform analysis based on discrete wavelet transform. For a given time series of packets arrivals, the LDE produces a logarithm plot of the data wavelet spectrum estimates. Although the LDE has the ability to identify correlation structures in the data trace [12], we mainly use it to assess the accuracy of the synthetic traces produced by LiTGen.

We first focus on web, mail and P2P traffic and generate three independent synthetic traces using a simple version of our generator. With this so called *basic* LiTGen, all traffic entities are generated from renewal processes using the empirical distributions extracted from the captured trace. No other additional dependency is introduced between the random variables. The three synthetic traces are then merged into a single one and compared to the filtered captured traffic composed of the same three applications. Figure 2(a) shows the resulting LDE spectra. Clearly, the synthetic trace produced by basic LiTGen (thin curve) does not match the captured traffic spectrum (thick gray curve). This simple version of LiTGen's underlying model does not succeed in reproducing the captured traffic scaling structure with a good accuracy.

Previous studies (*e.g.* [9]) pointed out that a great part of the LDE energy was due to the organization of packets within flows. This leads to refine LiTGen's model by introducing a dependency between the arrival process of packets within an object and the corresponding object's size. Note that this dependency may reflect the impact of TCP on packets inter-arrival times in objects of different sizes. In this extension, referred to as *extended* LiTGen, the arrival of packets within objects is still modeled by renewal processes, but for an object

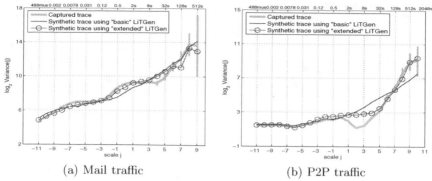

(a) Mail traffic   (b) P2P traffic

Fig. 3. Model evaluation: "basic" VS "extended" LiTGen in mail and P2P traffic

of a given size s, the inter-arrival random variables IA^s_{pkt} now depends on the object size. In order to evaluate extended LiTGen, we derive size-dependent empirical distributions of in-objects packets inter-arrivals, from the captured trace. When generating traffic, the packets inter-arrivals in an object of size s are taken from the corresponding IA^s_{pkt} distribution. The spectra obtained with extended LiTGen (circle curve in figure 2(a)) is barely distinguishable from the captured one. As a first result, the introduction of a simple dependency between the objects sizes and the packets inter-arrivals succeeds in reproducing accurately the traffic correlation structure, without taking into account network characteristics (such as TCP dynamics, RTT, loss rates). It thus appear that we do not need to introduce more complex non-renewal processes in the model, leading to a much simpler generator than the one developed in [4].

These three kinds of traffic, however, do not appear in the same proportions in the captured trace: while carrying 92.7% of the packets and 95.6% of the flows, web is the dominant application; mail carries 6.8% of the packets and 3.9% of the flows; P2P carries 0.5% of the packets and 0.5% of the flows. Figure 2(b) clearly indicates the differences between the three applications spectra in the captured trace that calls for studying each application independently. Figures 2(a) and 2(b) also show that our extended model accurately models the web traffic, conclusion reinforced by our previous study [8]. In the following, we thus focus on the mail and P2P traffics, which have been hidden by the predominant web traffic so far.

Figure 3(a) presents the mail traffic spectra. The reference spectrum (thick gray curve) corresponds to the captured mail traffic only. The basic LiTGen's underlying model (thin curve) reproduces in quite a good way the mail traffic spectrum. Extended LiTGen improves the results showing LiTGen's good ability to model mail traffic. Figure 3(b) shows the case of P2P traffic. Basic LiTGen fails to reproduce the captured traffic correlation structure for the scales above $j = 0$. The dip of reference spectrum at scales around $j = 2$ indicates a possible periodic behavior which we do not capture. Although the structural dependency introduced between N_{obj} and IA_{pkt} in extended LiTGen does not lead to the same improvement when dealing with P2P traffic, it allows the corresponding

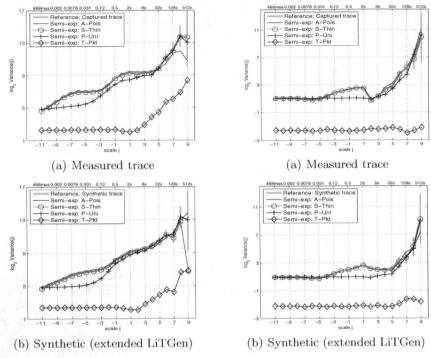

(a) Measured trace   (a) Measured trace

(b) Synthetic (extended LiTGen)   (b) Synthetic (extended LiTGen)

Fig. 4. Semi-experiments: mail trace   **Fig. 5.** Semi-experiments: P2P trace

spectra to match the reference one (except over scales comprised between $j = 0$ and $j = 5$). Due to space limitation, we do not provide here the investigation required to capture this apparent periodic behavior and leave it as future work.

While LiTGen exhibits good results on the overall spectra, we need a further advanced methodology to validate the internal properties of the synthetic traffic.

3.2   Semi-experiments Method

Semi-experiments have been introduced in [13] and consist in an arbitrary but insightful manipulation of internal parameters of the time series studied. The comparison of the energy spectrum before and after the semi-experiment leads to conclusions about the importance of the role played by the parameters modified by the semi-experiment. We apply the same set of semi-experiments to the captured traces and the synthetic traces generated by extended LiTGen. We then compare the impact of the internal manipulations to the two time series for the mail (Fig.4) and the P2P (Fig.5) traffic.

T-Pkt is a **T**runcation manipulation that allows to examine the objects arrival process by keeping only the first packet of each object. Removing packets decreases the energy of the spectrum that takes smaller values. As shown in figures 4 and 5, **T-Pkt** has a similar impact on the captured and the synthetic traces, for both mail and P2P traffic.

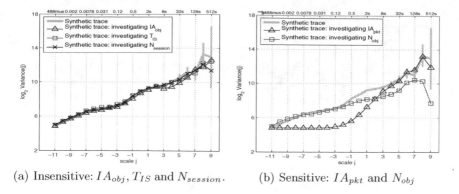

(a) Insensitive: IA_{obj}, T_{IS} and $N_{session}$.   (b) Sensitive: IA_{pkt} and N_{obj}

Fig. 6. Mail: test for the memoryless hypothesis

The **S-Thin** manipulation allows to test for the independence of objects. It randomly **S**elects objects with some probability, here equal to 0.9. When applying **S-Thin**, the spectra of the captured and synthetic trace, for the mail as well as P2P traffic, keep the same shape but drop by a small amount close to $log_2(0.9) = -0.15$.

A-Pois allows to examine the interactions between objects. This manipulation repositions the objects **A**rrival times according to a Poisson process and randomly permutes the objects order (while preserving the internal packet structure of objects). While **A-Pois** is a drastic manipulation, it has very little (and similar) effect on the spectra of all traces, indicating the negligible contribution of object arrival process in comparison to packets arrival.

P-Uni confirms this conclusion since it allows to examine the impact of in-objects packets burstiness. **P-Uni** uniformly distributes arrival times of packets in each object while preserving packets count and object duration. This manipulation flattens the spectrum from scales $j = -11$ to $j = -5$ for mail (resp. from scales $j = -11$ to $j = 2$ for P2P) in a comparable manner for the captured and synthetic traces.

As a conclusion, the captured and synthetic traces spectra present similar reactions to each semi-experiment manipulation. This indicates that LiTGen captured the key internal properties of the traffic highlighted by the semi-experiments, *i.e.* the object arrival process has few influence on the traffic burstiness; the objects can be considered as independent and the packets arrival process within objects contributes mostly to the energy spectrum. Note that the simple structure of our traffic description, which still relies on renewal processes, is sufficient to reproduce these traffic internal properties.

4   Impact of Traffic Entities Properties

We first investigate if "well-known" distributions can accurately approximate the empirical ones. To this aim, we use statistical quality of fit tests (*e.g.* KS-test)

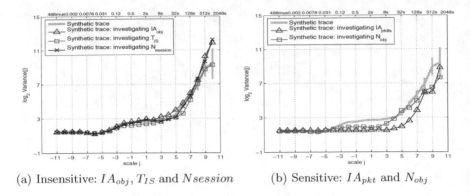

(a) Insensitive: IA_{obj}, T_{IS} and $Nsession$ (b) Sensitive: IA_{pkt} and N_{obj}

Fig. 7. P2P: test for the memoryless hypothesis

and compute indices of goodness of fit (*e.g.* Sum of Squares due to Error, R-Square...) to determine the "best" approximation. We led to similar conclusions for mail and P2P traffic. First, heavy-tailed distributions approximate well the random variables $N_{session}$ and N_{obj} (Pareto distributions) as well as IA_{obj} (Weibull distributions) and IA_{pkt} (close to lognormal distributions). Then exponential distributions approximate well T_{IS}.

The flexibility of extended LiTGen enables us to investigate the sensitivity of the traffic correlation structure with regards to the random variable distributions. To this aim, we replace individually the experimental distribution of each random variable by a memoryless distribution (exponential or geometric) of same mean. We thus create five synthetic traces, each one corresponding to a given random variable ($N_{session}$, T_{IS}, N_{obj}, IA_{obj} and IA_{pkt}). We then compare these traces to the reference synthetic trace generated by extended LiTGen calibrated with the experimental distributions.

Observing first the mail traffic, figure 6(a) shows that modeling the random variables IA_{obj}, T_{IS} and $N_{session}$ by memoryless distributions has a very small impact on the spectra of the LDE. On the contrary, modeling IA_{pkt} and N_{obj} by memoryless distributions widely impacts the spectra, as shown in figure 6(b). As an example, modeling IA_{pkt} by an exponential distribution completely erases the correlation existing between packets inter-arrivals and flattens the spectrum at scales below $j = -3$. This confirms the results obtained by the semi-experiments methodology that designated the in-object packets inter-arrival structure as the main source of energy in the spectrum. As shown in figure 7 similar conclusions can be drawn for P2P traffic.

For both P2P and mail, we observe that modeling IA_{obj}, T_{is} and $N_{session}$ by memoryless distributions is a realistic assumption that leads to an extremely small loss of accuracy compared to the reference spectrum. On the contrary, modeling IA_{pkt} and N_{obj} by memoryless distributions is far from realistic and indicates the need to model these random variables more carefully.

This investigation leads then to very interesting results illustrated by the number of objects within sessions $N_{session}$ and their arrival process IA_{obj}. Although the experimental distributions of these random variables are closely approximated by heavy-tailed distributions, we show that both distributions have negligible influence on the scaling behaviors in traffic. The presence of heavy-tailed distributions does not compulsorily imply a presupposed scaling behavior.

Nevertheless, the internal structure of objects has a strong influence on the spectra. In both mail and P2P traffic, the investigation concerning the packets inter-arrival distribution clearly points it out as the source of correlation in traffic at small scales.

5   Conclusion

This paper describes LiTGen, a per-user oriented traffic generator. LiTGen has the benefit to reproduce accurately the traffic scaling properties at small and large time scales, while using a very simple underlying hierarchical model. Thanks to LiTGen, we investigated the impact of the random variables distributions describing the IP traffic structure. This investigation is important for two reasons. First it helps understanding the sensitivity of the traffic, with regards to the distributions involved in its description, and then identify crucial parameters. It also gives insights to anyone willing to provide accurate traffic models. Most analytical models rely on simple Markovian hypothesis and one must be careful about their impact. Whenever useful to improve accuracy, one should replace some of them (the proper ones) by more appropriate assumptions. As an example, the exact original wireless traffic spectrum, can not be reproduced without 1) taking into account the organization of packets within objects 2) the use of heavy-tailed distributions to model objects size (in number of packets) and respective packet inter-arrivals. Moreover, our study demonstrated that the presence of heavy-tailed distributions in traffic does not necessarily implies the correlation, some of them can be modeled by memoryless distributions without impacting the traffic scaling properties.

This study also demonstrated the ability of a hierarchical model to reproduce accurately the characteristics of classes of traffic. The exhibition of results corresponding to other datasets is part of our ongoing works. Precise classes of applications (e.g. web and mail together, the main contributors to wireless traffic) will be defined soon to specify the utilization domain of the hierarchical model. While the results of LiTGen are proper for the P2P traffic, a more accurate but simple model for P2P traffic and other classes of application is still to be defined. This will become particularly important when mobile users will massively adopt new services and decrease the domination of web application in the overall traffic.

References

1. Mah, B.A.: An empirical model of http network traffic. In: IEEE Infocom. (1997)
2. Barford, P., Crovella, M.: Generating representative web workloads for network and server performance evaluation. In: ACM Sigmetrics. (1998)
3. Sommers, J., Barford, P.: Self-configuring network traffic generation. In: ACM IMC. (2004)

4. Vishwanath, K.V., Vahdat, A.: Realistic and responsive network traffic generation. In: ACM Sigcomm. (2006)
5. Crovella, M., Bestavros, A.: Self-similarity in world wide web traffic: Evidence and possible causes. In: ACM Sigmetrics. (1996)
6. Willinger, W., Taqqu, M.S., Sherman, R., Wilson, D.V.: Self-Similarity throught high-variability: Statistical analysis of ethernet LAN traffic at the source level. In: ACM Sigcomm. (1995)
7. Misra, V., Gong, W.B.: A hierarchical model for teletraffic. In: IEEE CDC. (1998)
8. Rolland, C., Ridoux, J., Baynat, B.: Hierarchical models for different kinds of traffics on CDMA-1xRTT networks. Technical report, UPMC - Paris VI, LIP6/CNRS (2006) http://www-rp.lip6.fr/~rolland/techreport.pdf.
9. Ridoux, J., Nucci, A., Veitch, D.: Seeing the difference in IP traffic: Wireless versus wireline. In: IEEE Infocom. (2006)
10. Donelson-Smith, F., Hernandez-Campos, F., Jeffay, K., Ott, D.: What TCP/IP protocol headers can tell us about the web. In: ACM Sigmetrics. (2001)
11. Veitch, D. and Abry, P.: Matlab code for the wavelet based analysis of scaling processes, http://www.cubinlab.ee.mu.oz.au/~darryl/.
12. Abry, P., Taqqu, M.S., Flandrin, P., Veitch, D.: Wavelets for the analysis, estimation, and synthesis of scaling data. In: Self-Similar Network Traffic and Performance Evaluation. Wiley (2000)
13. Hohn, N., Veitch, D., Abry, P.: Does fractal scaling at the IP level depend on TCP flow arrival process? In: ACM IMC. (2002)

Verification of Common 802.11 MAC Model Assumptions

David Malone, Ian Dangerfield, and Doug Leith

Hamilton Institute, NUI Maynooth, Ireland*

Abstract. There has been considerable success in analytic modeling of the 802.11 MAC layer. These models are based on a number of fundamental assumptions. In this paper we attempt to verify these assumptions by taking careful measurements using an 802.11e testbed with commodity hardware. We show that the assumptions do not always hold but our measurements offer insight as to why the models may still produce good predictions. To our knowledge, this is the first in-detail attempt to compare 802.11 models and their assumptions with experimental measurements from an 802.11 testbed. The measurements collect also allow us to test if the basic MAC operation adhere to the 802.11 standards.

1 Introduction

The analysis of the 802.11 CSMA/CA contention mechanism has generated a considerable literature. Two particularly successful lines of enquiry are the use of pure p-persistent modeling (e.g. [3]) and the per-station Markov chain technique (e.g. [2]). Modeling usually involves some assumptions, and in this respect models of 802.11 are no different. Both these models assume that transmission opportunities occur at a set of discrete times. These discrete times correspond to the contention counter decrements of the stations, equivalent to state transitions in the models, and result in an effective slotting of time. Note that this slotting based on MAC state transitions is different from the time slotting used by the PHY. A second assumption of these models is that to a station observing the wireless medium, every slot is equally likely to herald the beginning of a transmission by one or more other stations. In the models this usually manifests itself as a constant transmission or collision probability.

In this paper we will show detailed measurements collected from an experimental testbed to study these assumptions. This is with a view to understanding the nature of the predictive power of these models and to inform future modeling efforts. The contribution of this paper includes the first published measurements of packet collision probabilities from an experimental testbed and their comparison with model predictions and the first detailed comparison of measured and predicted throughputs over a range of conditions.

We are not the first to consider the impact of model assumptions. In particular, the modeling of 802.11e has required the special treatment of slots immediately

---

* This work was supported by Science Foundation Ireland grant IN3/03/I346.

S. Uhlig, K. Papagiannaki, and O. Bonaventure (Eds.): PAM 2007, LNCS 4427, pp. 63–72, 2007.
© Springer-Verlag Berlin Heidelberg 2007

after a transmission in order to accommodate differentiation based on AIFS (e.g. [1,9,11,6,4]). In [13] the nonuniform nature of slots is used to motivate an 802.11e model that moves away from these assumptions.

2   Test Bed Setup

The 802.11e wireless testbed is configured in infrastructure mode. It consists of a desktop PC acting as an access point, 18 PC-based embedded Linux boxes based on the Soekris net4801 [7] and one desktop PC acting as client stations. The PC acting as a client records delay measurements and retry attempts for each of its packets, but otherwise behaves as an ordinary client station. All systems are equipped with an Atheros AR5215 802.11b/g PCI card with an external antenna. All stations, including the AP, use a Linux 2.6.8.1 kernel and a version of the MADWiFi [8] wireless driver modified to allow us to adjust the 802.11e CWmin, AIFS and TXOP parameters. All of the systems are also equipped with a 100Mbps wired Ethernet port, which is used for control of the testbed from a PC. Specific vendor features on the wireless card, such as turbo mode, are disabled. All of the tests are performed using the 802.11b physical maximal data transmission rate of 11Mbps with RTS/CTS disabled and the channel number explicitly set. Since the wireless stations are based on low power embedded systems, we have tested these wireless stations to confirm that the hardware performance (especially the CPU) is not a bottleneck for wireless transmissions at the 11Mbps PHY rate used. As noted above, a desktop PC is used as a client to record the per-packet measurements, including numbers of retries and MAC-level service time. A PC is used to ensure that there is ample disk space, RAM and CPU resources available so that collection of statistics not impact on the transmission of packets.

Several software tools are used within the testbed to generate network traffic and collect performance measurements. To generate wireless network traffic we use mgen. We will often use Poisson traffic, as many of the analytic models make independent or Markov assumptions about the system being analysed. While many different network monitoring programs and wireless sniffers exist, no single tool provides all of the functionality required and so we have used a number of common tools including tcpdump. Network management and control of traffic sources is carried out using ssh over the wired network.

3   Collision Probability and Packet Timing Measurement

Our testbed makes used of standard commodity hardware. In [5] we developed a measurement technique that only uses the clock on the sender, to avoid the need for synchronisation. By requesting an interrupt after each successful transmission we can determine the time that the ACK has been received. We may also record the time that the packet was added to the hardware queue, and by inverting the standard FIFO queueing recursion we can determine the time the MAC spent processing the packet. This process is illustrated in Figure 1. For

the measurements reported here, we have refined the technique described in [5]
by making use of a timer in the Atheros card that timestamps the moment completed transmit descriptors are DMAed to host memory. This allows us to avoid
inaccuracies caused by interrupt latency/jitter. As will be shown later, in this
way we are able to take measurements with microsecond-level timing accuracy.

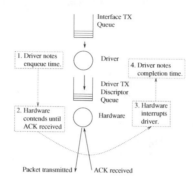

Fig. 1. Schematic of delay measurement technique

To measure packet collision probabilities, we make use of the fact that the
transmit descriptors also report the number of retry attempts R_i for each packet.
Using this we can estimate the calculate the total number of retries R and the
average collision probability $R/(P + R)$ where P is the number of successful
packet transmissions. We can also generalist this to get the collision probability
at the nth transmission attempt as

$$\frac{\#\,\{\text{packets with } R_i \geq n\}}{\#\,\{\text{packets with } R_i = n\} + \#\,\{\text{packets with } R_i \geq n\,\}}. \tag{1}$$

This assumes that retransmissions are only due to collisions and not due to
errors. We can estimate the error rate by measuring the retransmissions in a
network with one station. In the environment used, the error rate is $< 0.1\%$.

4   Validation

All the models we study assume that the 802.11 backoff procedure is being
correctly followed. The recent work of [12], demonstrates that some commercial
802.11 cards can be significantly in violation of the standards. In particular,
it has been shown that some cards do not use the correct range for choosing
backoffs or do not seem to back off at all. We therefore first verify that the cards
that we use perform basic backoffs correctly, looking at CWmin (the range of the
first backoff in slots), AIFS (how many slots to pause before the backoff counter
may be decremented) and TXOP (how long to transmit for).

To do this we measure the MAC access delay. This is the delay is associated
with the contention mechanism used in 802.11 WLANs. The MAC layer delay,

i.e. the delay from a packet becoming eligible for transmission (reaching the head of the hardware interface queue) to final successful transmission, can range from a few hundred microseconds to hundreds of milliseconds, depending on network conditions. In contrast to [12], which makes use of custom hardware to perform measurements of access delay, here we exploit the fine grained timing information available using the measurement technique described in the previous section to make access delay measurements using only standard hardware.

To test the basic backoff behaviour of the cards, we transmitted packets from a single station with high-rate arrivals and observed the MAC access delay for each packet. Figure 2(a) shows a histogram of these times to a resolution of $1\mu s$ for over 900,000 packets. We can see 32 sharp peaks each separated by the slot time of $20\mu s$, representing a CWmin of 32. This gives us confidence that the card is not subject to the more serious problems outlined in [12].

There is jitter, either in the backoff process or in our measurement technique. However, we can test the hypothesis that this is a uniform distribution by binning the data into buckets around each of the 32 peaks and applying the chi-squared test. The resulting statistic is within the 5% level of significance.

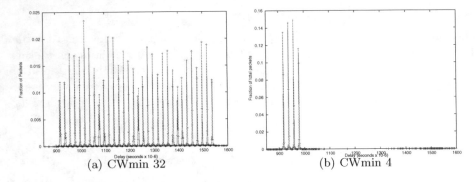

(a) CWmin 32   (b) CWmin 4

Fig. 2. Distribution of transmission times for packets with a single station. Note there a number peaks corresponding to CWmin.

The cards in question are 802.11e capable and so for comparison we adjust CWmin so that backoffs are chosen in the range 0 to 3. The results are shown in Figure 2(b) where we can see 4 clear peaks, as expected. We also see a small number of packets with longer transmission times. The number of these packets is close to the number of beacons that we expect to be transmitted during our measurements, so we believe that these are packets delayed by the transmission of a beacon frame.

Figure 3(a) shows the impact of increasing AIFS on MAC access time. In the simple situation of a single station, we expect increasing AIFS to increase MAC access times by the amount which AIFS is increased by. Comparing Figure 2(a) and Figure 3(a) confirms this.

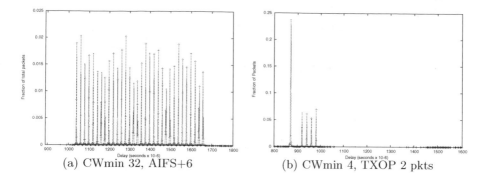

(a) CWmin 32, AIFS+6   (b) CWmin 4, TXOP 2 pkts

Fig. 3. Distribution of packet transmission times for a single station. On the left, AIFS has been increased, so the peaks are shifted by $120\mu s$. On the right, TXOP has been set to allow two packets to be transmitted every time a medium access is won, so we see approximately half the packets being transmitted in a shorter time.

Similarly, we can use TXOP on the cards to transmit bursts of packets, only the first of which must contend for channel access. Figure 3(b) shows the distribution of transmission times when two packet bursts are used. We see that half the packets are transmitted in a time almost $50\mu s$ shorter than the first peak shown in Figure 2(b).

These measurements indicate that a single card's timing is quite accurate and so capable of delivering transmissions timed to within slot boundaries. In this paper we do not verify if multiple cards synchronise sufficiently to fully validate the slotted time assumption.

5   Collision Probability vs Backoff Stage

Intuitively, the models that we are considering are similar to mean-field models in physics. A complex set of interactions are replaced with a single simple interaction that should approximate the system's behaviour. For example, by using a constant collision probability given by $p = 1 - (1 - \tau)^{n-1}$, where τ is the probability a station transmits, regardless of slot, backoff stage or other factors.

This assumption is particularly evident in models based on [2] as we see the same probability of collision used in the Markov chain at the end of each backoff stage. However similar assumptions are present in other models. It is the collision probability at the end of each backoff stage that we will consider in this section.

We might reasonably expect these sort of assumptions to better approximate the network when the number of stations is large. This is because the backoff stage of any one station is then a small part of the state of the network. Conversely, we expect that a network with only a small number stations may provide a challenge to the modeling assumptions.

Figure 4(a) shows measured collision probabilities for a station in a network of two stations. Each station has Poisson arrivals of packets at the same rate.

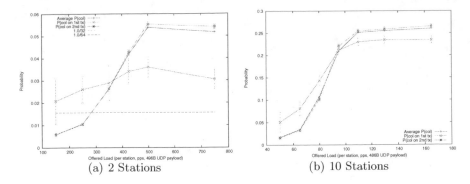

(a) 2 Stations   (b) 10 Stations

Fig. 4. Measured collision probabilities as offered load is varied. Measurements are shown of the average collision probability (the fraction of transmission attempts resulting in a collision), the first backoff stage collision probability (the fraction of first transmission attempts that result in a collision) and the second backoff stage collision probabilities (the fraction of second transmission attempts that result in a collision).

We show the probability of collision on any transmission, the probability of collision at the first backoff stage (i.e. the probability of a collision on the first transmission attempt for a given packet) and the probability of collision at the second backoff stage (i.e. the probability of collision at the second transmission attempt for a given packet, providing the first attempt was unsuccessful). Error bars are conservatively estimated for each probability using $1/\sqrt{N}$, where N is the number of events used to estimate the probability.

The first thing to note is that the overall collision probability is very close to the collision probability for the first backoff stage alone. This is because collisions are overwhelmingly at the first backoff stage: to have a collision at a subsequent stage a station must have a first collision and then a second collision, but we see that less than 4% of colliding packets have a collision at the second stage.

As we expect, both overall collision probability and first state collision probability increase as the offered load is increased. However, we observe that collisions at the second backoff stage show a different behaviour. Indeed, within the range of the error bars shown, this probability is nearly constant with offered load.

This difference in behaviour can be understood in terms of the close coupling of the two stations in the system. First consider the situation when the load is low. On a station's first attempt to transmit a packet, the other station is unlikely to have a packet to transmit and so the probability of collision is very low. Indeed, we would expect that the chance of collision to become almost zero as the arrival rate becomes zero.

Now consider the second backoff stage when the load is low. As we are beginning the second backoff attempt, the other station must have had a packet to transmit to have caused a collision in the first place. So, it is likely that both stations are on their second backoff stage. Two stations beginning a stage-two backoff at the same time will collide on their next transmission with probability $1/(2 * CWmin) = 1/64$ (marked on Figure 4(a)). If there is no collision, it is possible that the first station to transmit will have another packet available for transmission, and could collide on its next transmission, however as we are considering a low arrival rate, this should not be common.

On the other hand, if the load is heavy, it is highly likely that the other station has packets to send, regardless of backoff stage. This explains the increasing trend in all the collision probabilities shown. However, at the second backoff stage we know that both stations are have recently doubled their CW value. These larger than typical CW values result in smaller collision collision probability, and so we expect a lower collision rate on the second backoff stage compared to the first.

Figure 4(b) shows the same experiment, but now conducted with 10 stations in the network. Here, explicitly reasoning about the behaviour of the network is more difficult, but we see the same trends as for 2 stations: the first-stage and overall collision probabilities are very similar; collision probabilities increase as the load increases; collision probabilities at the second stage are higher than at first stage when the load is low, but vice versa when the load is high. The relative values of the collision probabilities are closer than in the case of 2 stations, but the error bars suggest they are still statistically different.

In contrast to the relatively gradual increase for two stations, we see a much sharper increase for 10 stations. Accurately capturing any sharp transition can be a challenge for a model.

In summary, while analytic models typically assume that the collision probability is the same for all backoff stages, our measurements indicate that this is generally not the case. However, collisions are dominated by collisions at the first backoff stage, and so the overall collision probability is a reasonable approximation to this. Adjustments to later-stage collision probabilities would represent second-order corrections when calculating mean-behaviour quantities (e.g. long term throughput). However, based on these measurements it is not clear if distributions or higher-order statistics, such as variances, predicted by existing models will always accurately reflect real networks.

6   Saturated Network Relationships

In this section we will consider the relationship between the average collision probability and the transmission probability. The relationship between these quantities plays a key role in many models, where it is assumed that

$$p = 1 - (1 - \tau)^{n-1}. \qquad (2)$$

Models will typically calculate τ based on mean backoff window or use a self-consistent approach, where a second relationship between p and τ gives a pair of equations that can be solved for both.

Once τ is known, the throughput of a system is usually calculated by calculating the the average time spent transmitting payload data in a slot by the average length of a slot. That is,

$$S = \frac{E_p n \tau (1 - \tau)^{n-1}}{\sigma (1 - \tau)^n + T_s n \tau (1 - \tau)^{n-1} + (1 - (1 - \tau)^n - n\tau (1 - \tau)^{n-1}) T_c}. \quad (3)$$

Here E_p is the time spent transmitting payload, σ is the time between counter decrements when the medium is idle, T_s is the time before a counter decrement after a successful transmission begins and T_c is the time before a counter decrement after a collision begins.

The pair of equations 2 and 3 are based on assuming that each station transmits independently in any slot. These equations can be tested independently of the rest of the model based on our measurements. Specifically, using our measurements of collision probability p, we may derive τ using equation 2 and then compare the predicted throughput given by equation 3 to the actual throughput.

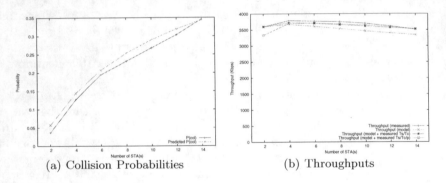

(a) Collision Probabilities   (b) Throughputs

Fig. 5. Predicted and measured collision probability (left) and throughput (right) in in a network of saturated stations as the number of stations is varied

Figure 5(a) shows the predictions made by a model described in [10] for the collision probabilities in a network of saturated stations and compares them to values measured in our testbed. We see that the model overestimates the collision probabilities by a few percent.

Figure 5(b) shows the corresponding measured throughput, together with model-based predictions of throughput made in several different ways. First, p and τ are predicted using the model described in [10] and throughput derived using equation 3. We take values $T_s = 907.8 \mu s$ and $T_c = 963.8 \mu s$ which would be valid if the 802.11b standard was followed exactly. It can be seen that, other than for very small numbers of stations, the model prediction consistently underestimates the throughput by around 10%.

Further investigation reveals that the value used for T_c appears to significantly overestimate the T_c value used in the hardware. While the standard requires that, following a collision, stations must pause for the length of time it would take to transmit an ACK at 1Mbps our measurements indicate that the hardware seems to resume the backoff procedure more quickly. In particular, values of $T_s = 916\mu s$ and $T_c = 677\mu s$ are estimated from test bed measurements. Using once again the model values for p and τ, but now plugging in our measured values for T_s and T_c, we see in Figure 5 that this produces significantly better throughput predictions, suggesting that the estimated values for T_s and T_c are probably closer to what is in use. In particular, we note that for larger numbers of nodes, where collisions are more common, the estimated throughput now closely matches the measured throughput.

Finally, instead of predicting p using a model, we use the measured value of p and estimate τ using equation 2. We continue to use the values of T_s and T_c based on testbed measurements. We can see from Figure 5 that for larger numbers of stations the throughput predictions are very similar to the previous situation. This suggests that equation 3 is rather insensitive to the small discrepancies seen in Figure 5 for larger numbers of stations. However, for two stations we see a significantly larger discrepancy in throughput prediction. This may indicate that the independence assumptions made by equations 2 and 3 are being strained by the strongly coupled nature of a network of two saturated stations.

7   Conclusion

In this paper we have investigated a number of common assumptions used in modeling 802.11 using an experimental testbed. We present the first published measurements of conditional packet collision probabilities from an experimental testbed and compare these with model assumptions. We also present one of the first detailed comparison of measured and predicted behaviour.

We find that collision probabilities are not constant when conditioned on a station's backoff stage. However, collisions are dominated by collisions at the first backoff stage, and so the overall collision probability is a reasonable approximation to this. Adjustments to later-stage collision probabilities would represent second-order corrections when calculating mean-behaviour quantities (e.g. long term throughput). However, based on these measurements it is not clear if distributions or higher-order statistics, such as variances, predicted by these models will always accurately reflect real networks.

We also find that throughput predictions are somewhat insensitive to small errors in predictions of collision probabilities when a moderate number of stations are in a saturated network. In all our tests, we see that two station networks pose a challenge to the modeling assumptions that we consider.

In future work we may explore the level of synchronisation between stations, the effect of more realistic traffic on the assumptions we have studied and the impact of non-fixed collision probabilities on other statistics, such as delay.

References

1. R Battiti and B Li. Supporting service differentiation with enhancements of the IEEE 802.11 MAC protocol: models and analysis. Technical Report DIT-03-024, University of Trento, 2003.
2. G Bianchi. Performance analysis of IEEE 802.11 Distributed Coordination Function. *IEEE JSAC*, 18(3):535–547, 2000.
3. F Cali, M Conti, and E Gregori. IEEE 802.11 wireless LAN: Capacity analysis and protocol enhancement. In *Proceedings of IEEE INFOCOM, San Francisco, USA*, pages 142–149, 1998.
4. P Clifford, K Duffy, J Foy, DJ Leith, and D Malone. Modeling 802.11e for data traffic parameter design. In *WiOpt*, 2006.
5. I Dangerfield, D Malone, and DJ Leith. Experimental evaluation of 802.11e edca for enhanced voice over wlan performance. In *International Workshop On Wireless Network Measurement (WiNMee)*, 2006.
6. P Engelstad and ON Østerbø. Queueing delay analysis of IEEE 802.11e EDCA. In *IFIP WONS*, 2006.
7. Soekris Engineering. http://www.soekris.com/.
8. Multiband Atheros Driver for WiFi (MADWiFi). http://sourceforge.net/projects/madwifi/. r1645 version.
9. Z Kong, DHK Tsang, B Bensaou, and D Gao. Performance analysis of IEEE 802.11e contention-based channel access. *IEEE JSAC*, 22(10):2095–2106, 2004.
10. D Malone, K Duffy, and DJ Leith. Modeling the 802.11 Distributed Coordination Function in non-saturated heterogeneous conditions. *To appear in IEEE ACM T NETWORK*, 2007.
11. JW Robinson and TS Randhawa. Saturation throughput analysis of IEEE 802.11e Enhanced Distributed Coordination Function. *IEEE JSAC*, 22(5):917–928, 2004.
12. A Di Stefano, G Terrazzino, L Scalia, I Tinnirello, G Bianchi, and C Giaconia. An experimental testbed and methodology for characterizing IEEE 802.11 network cards. In *International Symposium on a World of Wireless, Mobile and Multimedia Networks (WoWMoM)*, 2006.
13. Ilenia Tinnirello and Giuseppe Bianchi. On the accuracy of some common modeling assumptions for EDCA analysis. In *International Conference on Cybernetics and Information Technologies, Systems and Applications*, 2005.

Routing Stability in Static Wireless Mesh Networks

Krishna Ramachandran, Irfan Sheriff, Elizabeth Belding, and Kevin Almeroth

University of California, Santa Barbara

Abstract. Considerable research has focused on the design of routing protocols for wireless mesh networks. Yet, little is understood about the stability of routes in such networks. This understanding is important in the design of wireless routing protocols, and in network planning and management. In this paper, we present results from our measurement-based characterization of routing stability in two network deployments, the UCSB MeshNet and the MIT Roofnet. To conduct these case studies, we use detailed link quality information collected over several days from each of these networks[1]. Using this information, we investigate routing stability in terms of route-level characteristics, such as prevalence, persistence and flapping. Our key findings are the following: wireless routes are weakly dominated by a single route; dominant routes are extremely short-lived due to excessive route flapping; and simple stabilization techniques, such as hysteresis thresholds, can provide a significant improvement in route persistence.

1 Introduction

Applications, such as 'last-mile' Internet delivery, public safety, and distributed sensing, are driving the deployment of large-scale multi-hop wireless networks, also known as *mesh networks*. Although wireless routers in such networks are typically stationary, routes in these networks are expected to be unstable. One reason is that wireless links vary widely in their qualities because of multi-path fading effects, external interference and weather conditions. Link quality fluctuations can lead to variations in the quality of mesh routes, which can result in route fluctuations. This type of instability is unique to wireless networks.

Current routing protocols are not intelligent enough to consider routing stability during the selection of routes. A majority of the routing protocols [6] [14] ignore the fact that a route initially discovered has become sub-optimal over time. Route rediscovery is typically triggered by only route breaks and route timeouts. This approach can be detrimental to network performance.

Other routing protocols [2][7] periodically re-evaluate the quality of a route. The evaluation periodicity depends on the rate at which routing protocol control messages are exchanged. This approach fails to adapt to route quality variations that occur at smaller time-scales. However, by always picking the best route available, the resulting routing instability can lead to routing pathologies, such as packet reordering [3], which can severely degrade network performance.

---

[1] The collected datasets are available for download at
http://moment.cs.ucsb.edu/meshnet/datasets.

S. Uhlig, K. Papagiannaki, and O. Bonaventure (Eds.): PAM 2007, LNCS 4427, pp. 73–82, 2007.
© Springer-Verlag Berlin Heidelberg 2007

We require a routing protocol that provides the best tradeoff between performance adaptability and routing stability. A detailed investigation of routing stability can help us design such a routing protocol.

Another reason such an analysis is important is because routing stability impacts mesh network management. As an example, channel management schemes [15,16] in multi-radio mesh networks assign channels to frequency diversify routes in the mesh. If routes are expected to change, the mesh radios should also be re-assigned channels in order to ensure optimal network performance.

An understanding of routing stability can also help in network planning, such as router placement and radio configuration. For example, stability analysis may suggest that routes to certain regions in the coverage area fluctuate frequently. The reason could be either poor placement of routers or radio misconfiguration.

Although considerable research has focused on the design of routing protocols and routing metrics for wireless mesh networks, there exists no formal study of routing stability in such networks. This paper presents the first measurement-based characterization of routing stability in static wireless mesh networks. We perform our study by answering questions such as: (1) Is there a clear choice of an optimal route between a source-destination pair? (2) If not, how long do such routes persist before a route change (flap) occurs? (3) What benefit does a route flap provide? and (4) What measures can help reduce route flaps?

In order to perform our measurement-based characterization of routing stability, we analyze link-quality information collected over a period of 2-3 days from two mesh network deployments, the UCSB MeshNet[2], and the MIT Roofnet[3]. The MeshNet is a 20-node multi-radio 802.11a/b network deployed indoors on five floors of a typical office building on the UCSB campus. The MIT Roofnet is a 22-node outdoor network spread over four square kilometers in Cambridge, MA.

Clearly, routing stability analysis is influenced by the routing protocol. In order to investigate routing stability independent of any particular routing protocol, we compute high-throughput routes between all pairs of nodes assuming global knowledge of the collected link qualities. Routes are computed greedily, on a per-minute basis in our analysis, using the Dijkstra algorithm with the Weighted Cumulative Expected Transmission Time (WCETT) [7] as the path selection metric. We use WCETT because it has been shown to discover high throughput paths [7]. We compute routes greedily because we want to establish an upper bound on route capacities deliverable by a mesh network. Using the maximum capacities, we seek to understand the tradeoffs with respect to route instability.

The major findings from our study are as follows:

- Mesh routes are weakly dominated by a single route. The median prevalence of the dominant routes on the MeshNet and Roofnet are 65% and 57% respectively.
- Dominant routes are short-lived because of an excessive number of route flaps, most of which last only one minute.
- In a large number of cases, a route flap provides marginal improvement in throughput. 50% of the route flaps on the MeshNet, and 27% on the Roofnet, provide less than a 10% throughput improvement.

---

[2] http://moment.cs.ucsb.edu/meshnet
[3] http://pdos.csail.mit.edu/roofnet

– Avoidance of routes that either last only one minute or provide only 10% through-put improvement increases the lifetime of the dominant route up to five-fold on the MeshNet and up to four-fold on the Roofnet.

Although the above findings are specific to the two networks we have analyzed, we believe that the trends observed are generally applicable. Some of the findings discussed in this paper are well-known. A major contribution of this paper is a quantitative characterization of the extent of instability.

2 Related Work

Many studies have analyzed routing stability for wireline networks. Paxson reported on routing loops, routing stability, and routing symmetry by analyzing route information collected using *traceroute* [17]. Paxson found that Internet paths are typically dominated by a single route, and that a majority of Internet routes persist for either days or weeks. Labovitz et al. investigated Internet routing stability by analyzing BGP routing messages collected at key vantage points in the Internet [13]. Govindan et al. studied the growth of the Internet from 1994 to 1995 and found that route availability had degraded with the Internet's growth [9]. More recently, considerable attention has been given to routing pathologies because of BGP configuration faults [8,18].

In the domain of wireless networks, various routing protocols [2,6,14] have been proposed for multi-hop wireless networks. Although the discovery of routes has been extensively studied by these efforts, to the best of our knowledge, there exists no formal study of routing stability in such networks. Studies have investigated connectivity between source-destination pairs in mobile ad hoc networks in terms of the lifetime of routes [1]. However, in such networks, node mobility influences the route lifetime. Our focus is on static mesh networks where mobility has little bearing on routing stability. Instead, the stability is influenced by the network topology and variations in link quality.

3 Methodology

Our analysis of routing stability is based on link quality information collected from the UCSB MeshNet and the MIT Roofnet. We start this section by briefly describing the two deployments. We then discuss the technique used to collect link quality information, following which we present the route computation engine that uses the link qualities to compute routes. We end this section with a discussion of some shortcomings in our methodology.

3.1 Network Deployments

The UCSB MeshNet is a multi-radio 802.11a/b network consisting of 20 PC-nodes deployed indoors on five floors of a typical office building in the UCSB campus. Each node is equipped with two types of PCMCIA radios: a Winstron Atheros-chipset 802.11a radio and a Senao Prism2-chipset 802.11b radio. Each type of radio operates on a band-specific common channel. For rate adaptation, the 802.11b and 802.11a radios use auto-rate feedback [10] and SampleRate [2] respectively. There are 802.11b access points deployed in the building, which operate on various 802.11b channels. There is no external interference in the 802.11a band.

The MIT Roofnet consists of 22-nodes spread over four square kilometers in Cambridge, MA. Each node is a PC equipped with a Prism2-chipset 802.11b radio and an omni-directional antenna that is either roof-mounted or projecting out of a window. All radios operate on the same 802.11b channel. The Roofnet nodes experience interference from other, non-Roofnet access points.

3.2   Link Quality Estimation

Link quality is measured using the Expected Transmission Time (ETT) metric [7], which estimates the total time to transmit a packet on a link. The ETT is calculated from a link's loss rate and its data rate. ETT is given by the equation: $[(packetsize)/(d_1 * d_2 * bw)]$, where d_1 and d_2 are the link's delivery ratios in the forward and reverse directions, and bw is the average of the link data rate reported by the two end nodes on the link. $packetsize$ is assumed to be 1500 bytes.

In the case of the MeshNet, the link quality information was collected on three different days. The loss rate was calculated by having each node issue a broadcast probe of size 524 bytes every second on each of its radios. Each node records the number of probes received from each of its neighbors in a 10 second window. The ratio of the number of packets received to the number of packets sent (10) yields a link's delivery ratio. The link data rate is measured using packet pair probing [11]. Every 10 seconds, each node issues packet-pair unicast probes of size 134 bytes and 1134 bytes on each of its radios. The difference in transmission time of the packet pair, as measured by a neighbor, is piggybacked on packet pairs issued by that neighbor. Every 10 seconds, each node reports each of its link's delivery ratio and data rate to a central repository.

In the case of the Roofnet, link delivery ratios are available[4] on a per-minute basis for each 802.11b data rate. Since bandwidth information is not available for ETT computation, we set the link's ETT to be the ETT at the lowest data rate. In order to compute link delivery ratios, every 3 seconds, each Roofnet node broadcasts a 1500 byte probe at each of the 802.11b data rates, and a 134 byte probe at 1 Mbps. The 1500 byte probe is used to estimate the delivery probability of a large data packet at each of 802.11b data rates, whereas the 134 byte probe is used to estimate the delivery probability of a 802.11b acknowledgment. We use link delivery ratios on the 12th and 13th of May 2004 in our analysis.

3.3   Route Computation

We compute routes between all source-destination pairs for each minute recorded in our two data sets using an implementation of the Dijkstra's shortest-path algorithm. The quality of a route is computed using the Weighted Cumulative Expected Transmission Time (WCETT) metric [7]. The WCETT of a route is an estimate of the time a packet will take to traverse that route. The estimate is computed by taking into account the data

---

[4] http://pdos.csail.mit.edu/roofnet

rates, reliabilities, and channel assignments of all links on the path. We set WCETT's channel diversification parameter to 0.5. This setting gives equal weight to a path's channel diversification and its packet delivery rate [7]. In the case of the Roofnet, all radios operated on a common channel. Hence, channel diversification did not play a role in the route computation for the Roofnet. A total of 6,345 and 11,470 unique routes were observed for the MeshNet and the Roofnet, respectively.

3.4 Shortcomings

Some noteworthy shortcomings in our analysis methodology are worth considering. First, we do not explicitly account for the impact of network load and external networks on the link quality measurements. In the case of the UCSB MeshNet, there was no data traffic on the mesh during the collection period. We are unable to say for a fact that this was the case with the MIT Roofnet because the Roofnet was operational during the link quality monitoring. Both networks experienced interference on the 802.11b band. We believe that the outcome of our analysis does not change per se. However, with our current methodology, we are unable to quantify the extent of the impact of these factors on our results. We plan to address this shortcoming in our future work.

A second consideration is the relationship between routing stability and time-of-day patterns. Routing behavior is expected to be more stable during off-peak hours when external interference and the load on the network are typically low. Our current analysis does not differentiate routing behavior based on time-of-day patterns. We plan to investigate this effect in our future work.

Finally, the configuration of a radio, such as its transmission power, receive sensitivity, and carrier sense threshold, is likely to influence routing stability. A majority of current radios and their drivers do not permit fine-grained control of configuration settings. As a result, an empirical-based analysis of the impact of radio configuration on routing stability is challenging. Software-defined radios are likely to help address this limitation.

4 Stability Analysis

We use three stability metrics in our analysis. First, *prevalence* is the probability of observing a given route [17]. Second, *persistence* represents the duration for which a route lasts before a route change occurs [17]. Third, *route flap* refers to a change in route.

4.1 Route Prevalence and Persistence

For a given source-destination pair, we analyze its routing prevalence in terms of its dominant route. The dominant route is the route observed the most number of times. In order to compute p_d, the prevalence of the dominant route, we note n_p, the total number of times any route was available between the given pair as is observed in the set of routes computed using the technique described in Section 3.3; and k_p, the number of times the dominant route was observed in the same route set. The prevalence p_d is then given as $p_d = k_p/n_p$.

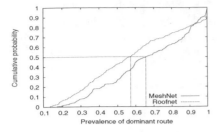

Fig. 1. Prevalence of the dominant route for all source-destination pairs

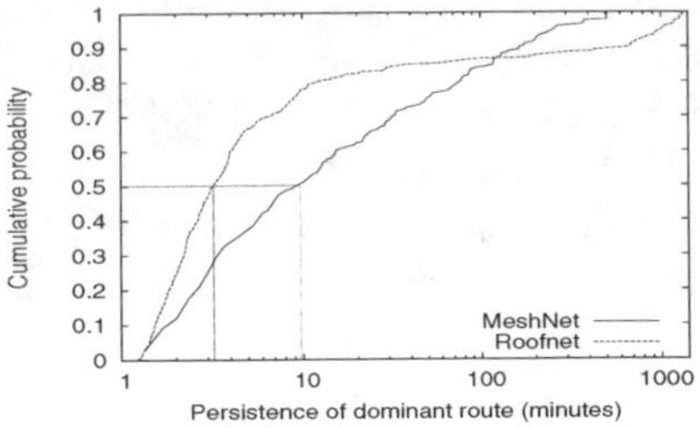

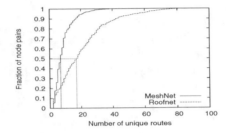

Fig. 2. Persistence of the dominant routes between all source-destination pairs

Fig. 3. Number of unique routes for all source-destination pairs

Figure 1 shows the cumulative distribution of the prevalence of the dominant route for all source-destination pairs in the MeshNet and Roofnet. We observe that the dominant routes in both networks have a wide distribution of prevalence values. The median prevalence on the MeshNet and Roofnet are 65% and 57%, respectively. This observation suggests that *routes in static mesh networks are weakly dominated by a single route*.

We next analyze the persistence of the dominant routes. In order to calculate the persistence of the dominant route, we record all the durations observed for each dominant route. The persistence of a dominant route is then computed as the average of all its recorded durations.

Figure 2 plots the cumulative distribution of the persistence values in minutes for the dominant routes. For better clarity, only persistence values in the range of 1-1200 minutes are depicted on the x-axis. We observe that the dominant routes for both networks have a wide distribution of persistence values. The median persistence value for the MeshNet is 9.6 minutes, and the corresponding value for the Roofnet is 3.2 minutes. This result suggests that *routes in static mesh networks are short-lived*.

Note that, in general, the prevalence and persistence of the dominant route in the MeshNet are higher than in the Roofnet. To investigate the reason, we examined the number of unique routes computed between all pairs of nodes in the two networks. Figure 3 shows the cumulative distribution of the number of unique routes for all source-destination pairs. For the median node pair, the MeshNet offers 7 unique routes while the Roofnet offers as many as 17 unique routes. In general, the number of unique routes available between node pairs in the Roofnet is much higher than in the MeshNet. Therefore, there exists a higher probability for a Roofnet node-pair to choose a route other than the dominant route, compared to a MeshNet node-pair. This reason could explain the lower prevalence and persistence values in the Roofnet compared to the MeshNet.

One plausible explanation for the higher number of available routes in the Roofnet lies in the difference in the design of the two networks. The Roofnet is an outdoor 802.11b network, whereas the MeshNet is an indoor 802.11a/b network. In spite of being a dual-radio mesh, we observed that the majority of routes in the MeshNet consisted of 802.11a links. This majority occurs because 802.11a offers significantly higher data rates as compared to 802.11b. Now, 802.11b has a greater range than 802.11a. 802.11a range is further limited in a non-line-of-sight indoor environment as is the case in the MeshNet. Consequently, the Roofnet nodes are better connected with one another than nodes in the Meshnet. This reason could explain why the number of routes available in the Roofnet is much higher than in the MeshNet.

A worthwhile consideration following from the above reasoning is the impact network planning has on routing stability. In the specific case of the MeshNet, network connectivity likely contributed to higher persistence and prevalence values compared to the Roofnet. As another case in point, Camp et al. found that node placement in their Houston urban mesh deployment influenced routing performance [4].

Our analysis of persistence and prevalence indicates that routes in wireless mesh networks are inherently unstable. As a result, one would expect route flaps to occur frequently in a mesh network. The next section investigates the utility of the route flaps by investigating the throughput improvement they offer, and their lifetimes.

4.2   Route Flapping

The methodology to analyze the impact of route flaps is as follows. Every route change between a source-destination pair from one instance of time to the next is recorded as a route flap. For each route flap, we noted the length of time, in minutes, the flap persists before the next flap is observed. Also, for each route flap, we computed the percentage throughput improvement offered by the new route over the old route. Assuming a 1500 byte packet, the throughput of a route can be computed by taking the ratio of packet size to the route's WCETT value.

Figures 4 and 5 plots the percentage throughput improvement offered by a route flap on the y-axis against the lifetime of the flap on the x-axis. Each point corresponds to a route flap. For better clarity, only flap lifetimes in the range 1 through 50 are depicted on the x-axis. Several observations can be made from this figure.

First, the figure shows a high concentration of short-lived route flaps. The long-lived flaps are smaller in number and likely correspond to the dominant routes. Figure 6 plots all the route flaps shown in Figures 4 and 5 as a cumulative distribution of their flap lifetimes. For both networks, over 60% of the route flaps last only a minute; 90% of the route flaps last less than five minutes. The high number of short-lived route flaps contribute to the instability of routing in the two networks, as is observed in our analysis in Section 4.1.

Second, even though a high concentration of short-lived route flaps exists, the throughput improvement offered by these flaps varies widely. For example, in both networks, the one minute route flaps offer throughput improvements as little as 0.001% and as high as 100,000%. The implication of our findings is that *opportunistic*

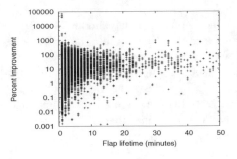

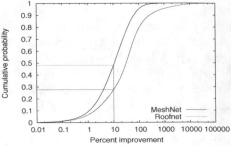

Fig. 4. Throughput benefit of route flaps in MeshNet

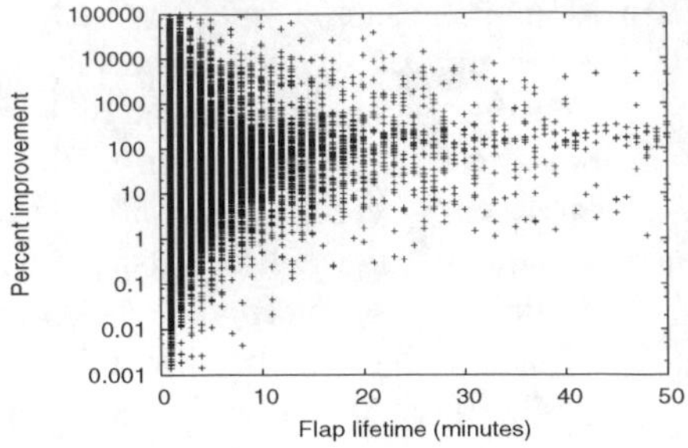

Fig. 5. Throughput benefit of route flaps in Roofnet

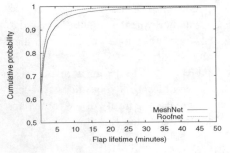

Fig. 6. Flap lifetimes as a fraction of total routes

Fig. 7. Percentage throughput improvement as a fraction of total routes

throughput maximization through route flaps can lead to significant instability in a mesh network. However, many short-lived routes do provide significant gains in throughput. This suggests a routing protocol that provides good stability may have to compromise on throughput gains.

A third observation is that a large number of route flaps provide only a marginal improvement in throughput. Figure 7 plots all the route flaps shown in Figures 4 and 5 as a cumulative distribution of the percentage throughput improvement they provide. 50% of the route flaps in the MeshNet and 27% of the route flaps in the Roofnet provide less than 10% throughput improvement. These route flaps vary in duration from 1 minute to 50 minutes. The implication of this result is that *a routing protocol that always flaps routes will likely achieve only minimal gains in a large number of instances.*

4.3   Can Routing Stability be Improved?

The previous observations suggest that route flapping can be dampened by selectively choosing an alternate route between a source-destination pair. For example, a routing protocol may choose to switch to an alternate route only when the route offers more than 10% throughput improvement over what is currently used. In the specific case of the UCSB MeshNet, such a dampening threshold has the potential to eliminate more than 50% of all route flaps. Another likely dampening metric could be to switch to an

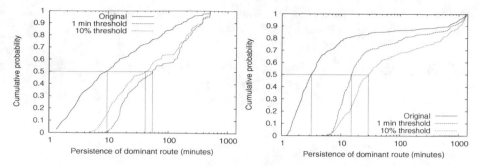

Fig. 8. Route stability with dampening in Mesh- **Fig. 9.** Route stability with dampening in
Net   Roofnet

alternate route only when the alternative is consistently better than the current route
for a specified amount of time. For example, this period could be two minutes. In the
specific case of the UCSB MeshNet, such a dampening strategy has the potential to
eliminate more than 60% of all route flapping.

To investigate the routing stability improvements that can result by applying such
dampening techniques, we use two dampening metrics. The first metric is a 10% throughput improvement threshold, i.e., an alternate route is chosen only if it provides better than
10% throughput improvement. The second dampening metric is an alternate route persistence value of two minutes, i.e., the alternate route is available for at least 2 minutes.

Figures 8 and 9 plots the results from our application of the dampening techniques.
The graphs depict the persistence values of the dominant routes against the fraction of
all dominant routes. In the case of the MeshNet, if we consider the median dominant
route, the one minute dampening metric yields a 5-fold increase in persistence. The
10% threshold yields a 4.5-fold increase in persistence. In the case of the Roofnet, the
10% threshold yields a 4-fold increase in persistence whereas the one minute threshold
yields a 3-fold increase.

The above results indicate that by *using low thresholds during route selection in a
mesh network, the persistence of the dominant routes can be significantly increased,
therefore leading to increased stability*. An increase in the persistence will reduce routing pathologies, such as packet reordering [3], but may lower end-to-end throughput.
As future work, we plan to investigate the trade-offs between stability and throughput
in more detail.

5   Conclusion

We present a measurement-based characterization of routing stability in two static wireless mesh networks. This is a first step towards understanding long term behavior of
routes in mesh networks. Some next steps for our continued analysis include: the impact of traffic load and external interference, the correlation between daily and weekly
patterns, and the impact of physical layer properties such as transmission power and
receiver sensitivity. We believe that the insights gained from this paper can stimulate

more research in understanding mesh routing behavior, which in turn can help us design better routing protocols and network management tools.

References

1. S. Agarwal, A. Ahuja, J. Singh, and R. Shorey. Route-lifetime Assessment Based Routing Protocol for Mobile Ad-hoc Networks. In *IEEE ICC*, New Orleans, LA, June 2000.
2. J. Bicket, D. Aguayo, S. Biswas, and R. Morris. Architecture and Evaluation of an Unplanned 802.11b Mesh Network. In *ACM MobiCom*, Cologne, Germany, August 2005.
3. E. Blanton and M. Allman. On Making TCP More Robust to Packet Reordering. *ACM Computer Communication Review*, 32(1):20–30, 2002.
4. J. Camp, J. Robinson, C. Steger, and E. Knightly. Measurement Driven Deployment of a Two-Tier Urban Mesh Access Network. In *ACM/USENIX Mobisys*, Uppsala, Sweden, June 2006.
5. A. Khanna, J. Zinky. The Revised ARPANET Routing Metric. In *ACM SIGCOMM*, Austin, TX, September 1989.
6. T. Clausen and P. Jacquet. Optimized Link State Routing Protocol. Internet Engineering Task Force, RFC 3626, October 2003.
7. R. Draves, J. Padhye, and B. Zill. Routing in Multi-radio, Multi-hop Wireless Mesh Networks. In *ACM MobiCom*, Philadelphia, PA, September 2004.
8. N. Feamster and H. Balakrishnan. Detecting BGP Configuration Faults with Static Analysis. In *USENIX Networked Systems Design and Implementation*, Boston, MA, May 2005.
9. R. Govindan and A. Reddy. An Analysis of Internet Inter-Domain Topology and Route Stability. In *IEEE Infocom*, Washington, DC, 1997.
10. A. Kamerman and L. Monteban. WaveLAN 2: A High-performance Wireless LAN for the Unlicensed Band. In *Bell Labs Technical Journal*, Summer 1997.
11. S. Keshav. A Control-Theoretic Approach to Flow Control. In *ACM Sigcomm*, Zurich, Switzerland, September 1991.
12. C. Labovitz, A. Ahuja, A. Bose, and F. Jahanian. Delayed Internet Routing Convergence. In *ACM Sigcomm*, Stockholm, Sweden, August 2000.
13. C. Labovitz, G. Malan, and F. Jahanian. Internet Routing Instability. *IEEE Transactions on Networking*, 6(5):515–528, 1998.
14. C. Perkins, E. Belding-Royer, and S. Das. Ad Hoc On-Demand Distance Vector Routing. Internet Engineering Task Force (IETF), RFC 3561, July 2003.
15. K. Ramachandran, E. Belding-Royer, K. Almeroth, and M. Buddhikot. Interference-Aware Channel Assignment in Multi-Radio Wireless Mesh Networks. In *IEEE Infocom*, Barcelona, Spain, April 2006.
16. A. Raniwala and T. Chiueh. Architecture and Algorithms for an IEEE 802.11-based Multi-Channel Wireless Mesh Network. In *IEEE Infocom*, Miami, FL, March 2005.
17. V. Paxson. End-to-end Routing Behavior in the Internet. In *ACM Sigcomm*, Palo Alto, CA, August 1996.
18. K. Varadhan, R. Govindan, and D. Estrin. Persistent Route Oscillations in Inter-domain Routing. *Computer Networks*, 32(1):1–16, January 2000.

Implications of Power Control in Wireless Networks: A Quantitative Study

Ioannis Broustis*, Jakob Eriksson, Srikanth V. Krishnamurthy, and Michalis Faloutsos

Department of Computer Science and Engineering,
University of California, Riverside
{broustis,jeriksson,krish,michalis}@cs.ucr.edu

Abstract. The use of power control in wireless networks can lead to two conflicting effects. An increase in the transmission power on a link may (i) improve the quality and thus the throughput on that link but, (ii) increase the levels of interference on other links. A decrease in the transmission power can have the opposite effects. Our primary goal in this work is to understand the implications of power control on interference and contention. We conduct experiments on an indoor mesh network. Based on analysis of our experimental data, we identify three interference scenarios: a) the overlapping case, where the aggregate throughput achievable with two overlapping links cannot be improved via power control; b) the hidden terminal case, where proper power control can primarily improve fairness and, c) the potentially disjoint case, where proper power control can enable simultaneous transmissions and thus improve throughput dramatically. We find that power control can significantly improve overall throughput as well as fairness. However, to our surprise, we note that using virtual carrier sensing in conjunction with power control generally degrades performance, often to a large degree.

Keywords: Wireless Networks, Interference, Carrier Sensing, Experimentation, Network Topology, Testbed.

1 Introduction

The goal of this paper is to characterize interference effects in IEEE 802.11-based wireless mesh networks and examine the impact of power control on interference. An increase in transmission power may result in: (a) increased quality of reception and hence, potentially higher throughputs at the intended receiver, and (b) increased interference levels. These two effects are conflicting in terms of providing the best network-wide throughput. We experimentally evaluate power control as a means of improving wireless network performance. We also evaluate the use of virtual carrier sensing (RTS/CTS) in conjunction with power control. In this study, we focus on the interference between pairs of links, and provide detailed experimental results for a wide variety of such pairs.

Based on our experimental results, we identify three interference scenarios. a) The *overlapping* case: if two links are overlapping, neither the use of power control nor the use of RTS/CTS messages can help improve the aggregate throughput. If two links

* This work is supported in part by the NSF CAREER Grant No. 0237920 and the NSF NRT grant No. 0335302.

S. Uhlig, K. Papagiannaki, and O. Bonaventure (Eds.): PAM 2007, LNCS 4427, pp. 83–93, 2007.

overlap, only one of the links can be active at any given time. Thus, the maximum total achievable throughput would be the maximum throughput achievable on one of these links. b) The *hidden-terminal* case: in this case, we find that proper power control is essential for ensuring fairness. To our surprise, we observe that the use of RTS/CTS in conjunction with power control generally results in degraded overall throughput and fairness in the hidden-terminal case. c) The *potentially disjoint* case: here, we find that power control can result in dramatic performance improvements. However, the use of virtual carrier sensing tends to remove opportunities for simultaneous transmissions, and consequently results in significantly lower overall throughput. We discuss these three cases in detail and explain the observed behavior.

Our work in perspective. Our results are based on measurements on an indoor testbed. Outdoor environments are subject to different constraints and may yield different results. We also limit our experiments to pairs of links. While we expect our results to carry over to multi-link interference, further studies are needed to verify this. While we do not propose an online power control algorithm, our results clearly demonstrate the need for an adaptive power control mechanism with the IEEE 802.11 MAC protocol. We believe that our study will stimulate further research on power control in a variety of settings, as well as provide insight towards the design of an adaptive power control mechanism. We wish to point out that while there are many studies on power control, they have been based predominantly on simulation or analysis. To the best of our knowledge, there have been no extensive experimentation studies on power control and interference in conjunction.

In section 2 we provide background on carrier sensing and discuss related studies on power control and interference. Section 3 describes our testbed and some initial experiments used for validation purposes. In section 4 we describe our experimental methods and present results related to interference effects. Section 5 concludes the paper.

2   Background and Previous Work

We revisit the IEEE 802.11 MAC protocol and in particular physical and virtual carrier sensing. We also discuss related studies and how our work differs from these efforts.

The Distributed Coordination Function (DCF) in the IEEE 802.11 MAC uses the CSMA/CA algorithm. In particular, this algorithm hinges on both physical and virtual carrier sensing. With physical carrier sensing, in order to avoid collisions, a node that intends to transmit first senses the medium. If an ongoing transmission is detected, the node tries again after a back-off period that is specified by the IEEE standard [2]. If no traffic is detected, the node proceeds with its transmission. However, physical carrier sensing suffers from the presence of *hidden* and *exposed* terminals. First, two transmitters may be unable to sense each others' carriers and hence, their transmissions may collide at the intended receiver. In this case, the two transmitters are said to be *hidden* from each other. Second, a transmitter (say A) might supress its transmission since it detects a carrier from another transmitter (say B); however, this may be overly conservative, as simultaneous transmissions are sometimes feasible despite A hearing B's carrier. In this case, transmitter A is said to be *exposed* to B's transmission. Virtual carrier sensing in IEEE 802.11, using RTS (Request-To-Send) and CTS (Clear-To-Send) control messages, is intended to address the hidden terminal problem, and has been known to exacerbate the exposed terminal problem. More details on the DCF function may be found in [2].

Next, we briefly discuss relevant previous work on interference characterization and power control in 802.11-based networks. As with our work, these efforts assume omnidirectional communications. Power control in mesh networks has received a lot of attention since it is attractive in terms of providing energy savings and spatial reuse. However, most efforts assume that the network can be modeled using unit disk graphs (UDGs) [16,6,17,11]. There have been more recent efforts that account for more realistic channel models [10,8]. However, most of these efforts rely on simulations and analyses; the models used may not be appropriate in all settings although they do provide better representations of the wireless channel. In [10], Muqattash and Krunz, design a power controlled MAC protocol that allows nodes to adjust their transmission powers while allowing for some interference margin at the receiver. In [7], distributed power control algorithms are designed; the algorithms take into account node sensitivities to current interference levels.

Experimental studies on wireless networks have recently gained popularity. Studies of wireless mesh networks in [5] and [4] demonstrate that the popular unit disk graph models are unlikely to hold in real networks. However, these efforts do not characterize interference between links nor do they consider power control.

Akella et al. [1] use large-scale measurement data from various cities to show how common it is to have tens of APs deployed in close proximity of each other. The paper proposes a power control and rate adaptation scheme (PERF) for the purpose of reducing interference among neighbors. We discuss PERF further in section 4. Padhye et al. [13] also study the problem of estimating pairwise interference among links in a multi-hop wireless testbed. They propose a link interference estimation method and study it in a variety of settings. Sinha et al., in [18], perform experiments to observe the differences between unicast and broadcast link properties. To study the impact of interference, they consider various distance-related interference scenarios on a grid indoor-testbed. Son et al. [15] study the effects of transmission power control on individual wireless link quality. They focus on sensor networks and perform their experiments on low-power RF links (up to 10 dBm). The interference effects are not studied.

3   Experimental Setup

In this section we describe our experimental testbed, and present initial experimental results on the stability and power controlled behavior for isolated links. These results provide a basis for our studies on the interference effects.

Our indoor testbed is comprised of 15 Soekris net4826 nodes [14], deployed in the 3rd floor of Engineering Building Unit II at the University of California, Riverside; the network is depicted in figure 1. Each node runs a Debian v3.1 Linux distribution with kernel version 2.6.13.2 and mounts its root partition over NFS from a server at start-up. We have equipped nodes with EMP-8602-6G 802.11a/b/g WiFi cards, which embed the Atheros AR5006 chipset [3]; the cards are controlled by the Linux MadWifi driver (version 0.90) [9]. Each card is connected to a 5-dBi gain, external omnidirectional antenna.

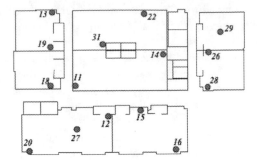

Fig. 1. Our indoor–testbed deployment. Nodes are represented by dots along with their IDs.

We use the 802.11a mode in order to avoid interference from co-located 802.11b/g networks; our testbed is the only 802.11a network in the area. For our experiments, we use channel 56, which corresponds to 5.28 GHz. The transmission power is varied between 1 and 16 dBm in our experiments. We use SampleRate [4,5] as the rate control algorithm. In order to derive reliable measurements, we run multiple experiments between many different pairs of links, and for different time periods of the day. We use 30 seconds of back-to-back 1500-byte UDP packets as our traffic load.

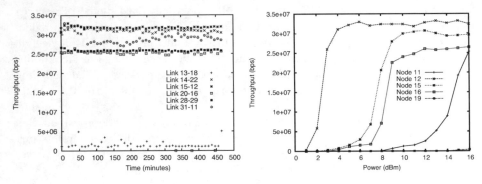

Fig. 2. Links exhibit relatively small variability over time

Fig. 3. Throughput (bps) vs. power (dBm) for the links of node 20

Reliability of Results. In order to determine the stability of our results, we ran a large number of 30-sec experiments, considering varying powers on a large set of links, each of which was activated in isolation. Figure 2 depicts the throughputs observed for 6 representative links, for 16 dBm of power. We observe that there is only a minor variation in performance, for most of the links under study. The throughput on the link 13→18 exhibits higher variance than the other links. This is because this link is relatively poor, as one can see from the achieved throughput. This set of experiments demonstrates the stability of our experimental results over time.

Throughput vs. Transmission Power. We observe the quality of isolated individual links (only one link is active at any given moment), for varying transmission power settings. As shown in figure 3, the throughput achieved on any given link depends on the transmission power used. We also observe that the maximum throughput for a link is achieved at a certain power threshold. For example, for the link 20→12 we observe that maximum throughput is achieved with about 4 dBm of transmission power. The throughput saturates here and a further increase in the transmission power does not significantly increase performance. Thus, isolated links do not always have to transmit with maximum power, in order to achieve maximum throughput. One may attribute this to the limited modulation and encoding possibilities with IEEE 802.11a.

4   Effect of Transmission Power on Interference and Contention

The overarching objective of our experimental study is to understand the transmission power trade-off: increased reliability and performance on one link vs. increased levels of interference caused to other links.

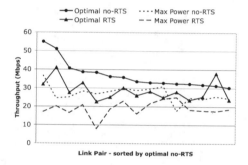

Fig. 4. Using the maximum transmission power does not result in the optimal throughput

In our experiments, we activate two links at a time. We then study the throughput achieved by each link as we vary their respective transmission powers. The transmitters of these links send bulk UDP traffic to their respective receivers. The transmission power is selected between 1 and 16 dBm, and ACKs are transmitted at the same power as DATA packets. Experiments have a 30-second duration, during which transmission power is maintained at a fixed level. After the expiration of this time period, each link is configured with a new power setting, and the experiment is repeated. We use the same carrier sensing threshold for all experiments (the default setting in our cards). We utilize the *iperf* measurement tool (version 2.0.2) [12] to measure link performance. We consider cases where we (i) disable or (ii) enable virtual carrier sensing with RTS/CTS.

4.1   Achievable Performance Gains

We would like to determine by how much, if at all, performance can be improved through the use of an optimal power control mechanism. Since such a power control mechanism is not available, we resort to exhaustive search. Figure 4 depicts the maximum total throughput observed for a set of link pairs when using the optimal power settings, as well as when using the maximum transmission power. Both the cases of having

88   I. Broustis et al.

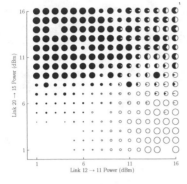

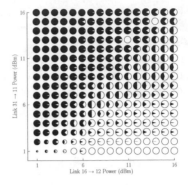

Fig. 5. A pair of overlapping links. Power control cannot improve the performance.

Fig. 6. A pair of links with a hidden terminal problem. Power control necessary for fairness.

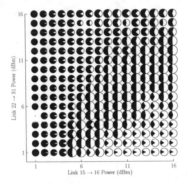

Fig. 7. A pair of potentially disjoint links. Power control can improve throughput up to 100%.

RTS/CTS enabled or disabled are plotted. We observe that using maximum power does not necessarily result in maximum network throughput. Thus, an opportunity exists for significantly improving network performance through power control. Moreover, using RTS/CTS does not appear to improve performance for any of the link pairs in question. We address this in more detail below. Next, we take a closer look at the collected data, to see the effect of transmission power settings on throughput as well as fairness.

4.2 Three Types of Interference Behavior

The above experiments were run on a set of 16 representative link pairs. Analyzing the experimental data in more detail, we observed that each of the 16 pairs could be categorized as belonging to one of three general cases. The three interference cases identified were **overlapping** (5 links out of 16), **hidden-terminal** (9 links out of 16) and **potentially disjoint** (2 links out of 16).

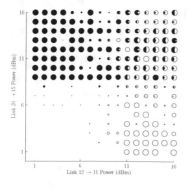

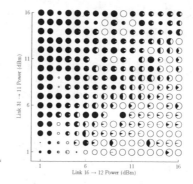

Fig. 8. Achieved throughputs for the overlapping case, when RTS/CTS is enabled

Fig. 9. RTS/CTS worsens performance and fairness in the hidden terminal case

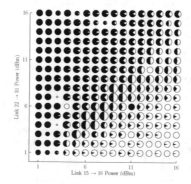

Fig. 10. Enabling RTS/CTS degrades performance in the disjoint case

We illustrate the three different types using the **throughput matrices** of representative link pairs (figures 5, 6 and 7). A throughput matrix illustrates the throughput achieved by each link, for all combinations of transmission powers. In figure 5, the horizontal and vertical axes represent the powers used by each link, respectively. The area of a disc corresponds to the sum total throughput achieved. The black and white portions of each disc correspond to the throughputs of the links indicated on the vertical and horizontal axes, respectively. For example, in figure 6, at coordinate (11,8), good throughput is achieved, and each link received exactly half of that throughput (17.5 Mbits/sec each). At coordinate (11,16), total throughput was again good, but link $31 \rightarrow 11$ received most of that throughput (30.4 Mbits/sec), leaving link $16 \rightarrow 12$ with very little (2.55 Mbits/sec). We will now describe each case in more detail, and study its effects in the corresponding throughput matrix.

Overlapping. In this case, the two links always contend, irrespective of the power settings used. Figure 5 shows an example of such a case. This case typically appears where the channel between the two senders is better than the channel from at least one sender to its corresponding receiver. In this example, nodes 12 and 20 are the transmitters and nodes 11 and 15 are the receivers, respectively (figure 1). The link 12→11 is a poor link and node 12 needs to transmit with at least 12 to 13 dBm to achieve a reasonable throughput. Similarly, node 20's power has to be at least 8 to 9 dBm. Transmitters 20 and 12 are within line-of-sight of each other (as seen in figure 1) and the link between them is very good. As a result, they can hear each others' transmissions and back-off accordingly. Hence, only one of the two links will carry traffic at a time, and the maximum total throughput of the two links will never be higher than the maximum of what is achievable on the best of the two links.

Hidden Terminal. The majority of the considered link pairs belong to this type. Here, senders cannot detect each others' transmissions and physical carrier sensing fails. Multiple transmissions arrive simultaneously at a receiver. This is not necessarily a problem if the signal strength of the desired transmission is significantly higher than that of the interfering transmission. This competition between signal and interference is observed in figure 6. As node 16 increases its power, it causes higher levels of interference on the link 31→11; collisions occur at node 11. However, as node 31 increases its transmission power, its signal strength increases. Hence node 11 is now able to decipher node 31's data packet. Note that while total throughput varies little with transmission power, if this power is over a certain level, fairness is consistently better along a diagonal in the throughput matrix. Power settings along this diagonal optimally balance signal quality and interference between the two links.

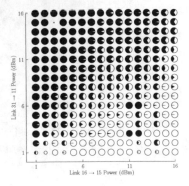 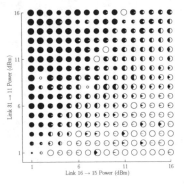

Fig. 11. Throughput matrix for a second link pair suffering from a hidden terminal problem, without RTS/CTS

Fig. 12. Throughput matrix for the second pair, with RTS/CTS. Though RTS/CTS in some cases improves fairness, this is at the cost of severely degraded overall throughput.

Potentially Disjoint. Potentially disjoint links are links on which, accurate power settings enable simultaneous successful transmissions, resulting in up to a doubling of the overall throughput. Potentially disjoint links are characterized by peaks in the throughput matrix exceeding the maximum capacity of a single link. This is depicted in figure 7, for the links 15→16 and 22→31. These two links are topologically well-separated (figure 1) and are of high quality (there is a line-of-sight path between the transmitters and their corresponding receivers). The maximum total throughput (approximately twice the capacity of a single link) is achieved when link 15→16 uses power 9 dBm and when link 22→31 uses 6 dBm. This suggests that when the transmission powers are set to the above values, the two links can operate interference-free. As links increase their powers, we observe that the total throughput decreases.

From these observations it is apparent that in order to achieve high total network throughput, appropriate setting of the transmission power is essential. In other words, an online power adaptation mechanism is needed that can, for each link, determine the power that is the most appropriate for use at a given instance in time. One proposed scheme, PERF [1], partially addresses this issue. PERF gradually decreases transmission power as long as throughput is not affected. PERF can be expected to successfully address the potentially disjoint case, correctly reducing power until spatial reuse is made possible. However in the hidden terminal case, PERF will use the maximum power, resulting in suboptimal fairness conditions. While PERF is a step in the right direction, we conclude that further improvements are necessary.

4.3   Use of RTS/CTS in Indoor Environments

Finally, we investigate how adding the RTS/CTS handshake affects interference and contention characteristics. We repeat the previous experiments, but with RTS/CTS enabled. Given that the purpose of RTS/CTS is to address the hidden terminal problem, we expect the results to somewhat improve in the hidden terminal cases. From the measurements we observe the following:

Overlapping. The performance of overlapping links is slightly worsened due to RTS/CTS overhead (figure 8).

Hidden Terminal. We were surprised to note that RTS/CTS consistently underperformed plain vanilla CSMA in *all* of the hidden terminal cases we tested. Figure 9 shows the throughput matrix for a typical hidden terminal case. Overall throughput is consistently lower when RTS/CTS is enabled. Fairness was affected to an even larger degree, to the extent that the stability of our results was significantly impacted. Note that there exist cases where, for some power settings, the use of RTS/CTS resulted in improved fairness. However, this improvement is generally offset by a large reduction in overall throughput (see figures 11, 12). We suspect that this is due to RTS/CTS exacerbating the exposed terminal problem, reducing the opportunity for parallel transmissions.

Potentially Disjoint. Finally, links in a potentially disjoint case are also negatively impacted by RTS/CTS (see figure 10). While overall throughput can still reach levels higher than the capacity of a single link, it is significantly lower than what was achieved with plain vanilla CSMA. The *regime* of powers in which spatial reuse is achieved is also reduced in size, requiring more precise power control, and generally higher transmission powers (compare to figure 7).

In conclusion, RTS/CTS appears to be entirely detrimental to the performance of indoor wireless networks, if power control is available. However, additional experiments need to be performed, in a wider range of environments and with a variety of hardware, before any firm conclusions be drawn.

5 Conclusion

While power control can affect the quality of transmission on a given link, the used power dictates the interference projected on other links and thus, can also affect the performance on other links. We perform measurements to quantify this trade-off with power control in an indoor experimental network with IEEE 802.11a nodes, both with and without virtual carrier sensing. We identify three types of interference behavior. In the *overlapping* case, power control and/or RTS/CTS do not increase the maximum achievable throughput. In the *hidden terminal* case, power control alleviates the hidden terminal problem and thus, improves the throughput. However, the use of RTS/CTS in conjunction with power control consistently degrades both the overall throughput and fairness. In the *potentially disjoint* case, power control can help in activating the links simultaneously and thereby can yield almost a twofold increase in the achievable throughput as compared to a case where the default maximum power is used. Again, virtual carrier sensing exacerbates the exposed terminal problem, significantly degrading overall performance.

We used IEEE 802.11a nodes, to avoid interference from co-located 802.11b/g networks. Even though results with 802.11b/g may differ for the links under investigation, our conclusions regarding the observed types of interference will still hold. Overall, our studies suggest that accurate power control holds great promise for improving the performance of indoor wireless networks.

References

1. A. Akella, G. Judd, S. Seshan, and P. Steenkiste. Self-management in chaotic wireless deployments. In *MOBICOM*, 2005.
2. ANSI/IEEE802.11-Standard. 1999 edition.
3. Atheros/AR5006chipset.
   http://www.atheros.com/pt/ar5006bulletins.htm.
4. J. Bicket. Bit-rate selection in wireless networks. In *Master's Thesis, Department of Electrical Engineering and Computer Science, MIT*, 2005.
5. J. Bicket, D. Aguayo, S. Biswas, and R. Morris. Architecture and evaluation of an unplanned 802.11b mesh network. In *MOBICOM*, 2005.
6. T. ElBatt, S. Krishnamurthy, D. Connors, and S. Dao. Power management for throughput enhancement in wireless ad-hoc networks. In *IEEE ICC*, 2000.
7. J. Huang, R. Berry, and M. Honig. Distributed interference compensation for wireless networks. In *IEEE JSAC, Vol. 24, No.5*, May 2006.
8. S. Kandukuri and S. Boyd. Optimal power control in interference-limited fading wireless channels with outage-probability specifications. In *IEEE Transactions of Wireless Communications, Vol. 1, issue 1, ISSN: 1536-1276, pp. 46-55*, Jan. 2005.
9. MadWifi-Driver. http://madwifi.org.
10. A. Muqattash and M. Krunz. Powmac: A single-channel power-control protocol for throughput enhancement in wireless ad hoc networks. In *IEEE JSAC, Vol. 23, No. 5, pp. 1067-1084,*, May 2005.
11. T. Nandagopal, T. Kim, X. Gao, and V. Bharghavan. Achieving mac layer fairness in wireless packet networks. In *MOBICOM*, 2000.
12. NLANR/iperf-version2.0.2. http://dast.nlanr.net/projects/iperf/.

13. J. Padhye, S. Agarwal, V. N. Padmanabhan, L. Qiu, A., and B. Zill. Estimation of link interference in static multi-hop wireless networks. In *IMC*, 2005.
14. Soekris/net4826. http://www.soekris.com/net4826.htm.
15. D. Son, B. Krishnamachari, and J. Heidemann. Experimental study of the effects of transmission power control and blacklisting in wireless sensor networks. In *IEEE SECON*, 2004.
16. R. Wattenhofer, L. Li, P. Bahl, and Y.-M. Wang. Distributed topology control for power efficient operation in multihop wireless ad hoc networks. In *INFOCOM*, 2001.
17. X. Yang and N. Vaidya. Priority scheduling in wireless ad hoc networks. In *MOBIHOC*, 2002.
18. H. Zhang, A. Arora, and P. Sinha. Learn on the fly: Beacon-free link estimation and routing in sensor network backbones. In *IEEE INFOCOM*, 2006.

TCP over CDMA2000 Networks: A Cross-Layer Measurement Study

Karim Mattar[1], Ashwin Sridharan[2], Hui Zang[2], Ibrahim Matta[1], and Azer Bestavros[1]

[1] Department of Computer Science at Boston University
{kmattar,matta,best}@cs.bu.edu
[2] Sprint Advanced Technology Labs
{ashwin.sridharan,hui.zang}@sprint.com

Abstract. Modern cellular channels in 3G networks incorporate sophisticated power control and dynamic rate adaptation which can have a significant impact on adaptive transport layer protocols, such as TCP. Though there exists studies that have evaluated the performance of TCP over such networks, they are based solely on observations at the transport layer and hence have no visibility into the impact of lower layer dynamics, which are a key characteristic of these networks. In this work, we present a detailed characterization of TCP behavior based on cross-layer measurement of transport, as well as *RF* and *MAC* layer parameters. In particular, through a series of active TCP/UDP experiments and measurement of the relevant variables at all three layers, we characterize both, the wireless scheduler in a commercial CDMA2000 network and its impact on TCP dynamics. Somewhat surprisingly, our findings indicate that the wireless scheduler is mostly insensitive to channel quality and sector load over short timescales and is mainly affected by the transport layer data rate. Furthermore, we empirically demonstrate the impact of the wireless scheduler on various TCP parameters such as the round trip time, throughput and packet loss rate.

1 Introduction

With advances in error-correction coding, processing power and cellular technology, the wireless channel need no longer be viewed as an error-prone channel with low bandwidth. Instead, modern 3G cellular networks (e.g CDMA2000 1xRTT, EV-DO, HSDPA/UMTS) deploy ARQ mechanisms for fast error recovery, as well as sophisticated wireless schedulers that can perform "on-the-fly" rate adaptation. The latter feature allows the network to adapt to diverse conditions such as channel quality, sector load and more importantly, as we show in this work, data backlog.

The *dynamic rate* adaptation of modern cellular channels implies that a source will typically experience variable bandwidth and delay, which may be caused by the scheduler's dependency on buffer backlog. Since TCP, the dominant transport protocol in the Internet, utilizes feedback from the channel to control its transmission rate (indirectly the buffer backlog), this creates a situation where two controllers, the wireless scheduler and TCP, share a single control variable.

There are several interesting studies that have considered the performance of TCP over cellular networks [2,5,9]. However, they mostly rely on measurement of TCP dynamics at the transport layer and have no visibility into the underlying MAC nor the

S. Uhlig, K. Papagiannaki, and O. Bonaventure (Eds.): PAM 2007, LNCS 4427, pp. 94–104, 2007.

dynamics of the radio channel. In this work, we measure relevant information at all three layers in a commercial CDMA2000 network to identify the dominant factors that affect TCP. To the best of our knowledge, this is the first study that looks at cross-layer measurements in a wireless network. Our contributions can be summarized as follows:

1. We conducted extensive active measurements in a commercial CDMA2000 cellular network to characterize the behavior of the wireless scheduler, and evaluate TCP's performance. One of our objectives was to identify the impact of various network factors on both the wireless scheduler and TCP. Towards this end, we develop a simple Information Theoretic framework that allows us to *quantify* how factors such as channel quality, sector load, *etc.*, affect the wireless scheduler, and how the scheduler in turn affects TCP.

2. In terms of the wireless scheduler, we exposed the different mechanisms that govern its operation and identified the characteristics that influence its performance. We concluded that over short timescales (1 second), the wireless scheduler: a) is highly dependent on buffer backlog, b) is surprisingly insensitive to variations in channel quality or sector load, and c) has a rate limiting mechanism to maintain fairness by throttling connections that are being persistently greedy. Over long timescales (20 minutes), however, the scheduler reduces allocated rate in response to persistently bad channel conditions or high sector load, and is unable to maintain fairness among concurrent TCP sessions.

3. In terms of TCP, we concluded that: a) there is a tight coupling between the TCP sending rate and the scheduler. This implies that rate variations, seen by TCP, on the CDMA channel are not random, b) most of the packet losses seen by TCP are congestion related, and c) the high variability in channel rate causes frequent spurious re-transmissions which can be overcome by using the time-stamp option.

4. Finally, as a general observation, we found high variability in TCP throughput based on the time and day of the experiment. We hypothesize this to be due to rapid cell dimensioning by network operators.

The rest of the paper is organized as follows: Section 2 outlines the architecture of a CDMA2000 network and highlights the relevant features. Section 3 presents a description of the various experiments that we conducted. Section 4 characterizes the wireless scheduler and quantifies the relative impact of various factors on it. Section 5 presents an evaluation of TCP's performance. Section 6 presents our conclusions.

Due to lack of space, some characterization details were omitted from this paper. The reader is referred to [7] for all the details.

2   The CDMA2000 1xRTT System

Figure 1(a) depicts the architecture of a typical cellular data network in order to illustrate its salient features. The network consists of two main components: a) the *data network* which is responsible for operations like managing PPP sessions, IP mobility and billing, and b) the *radio network* which manages radio resources. The focus of this work is on the latter.

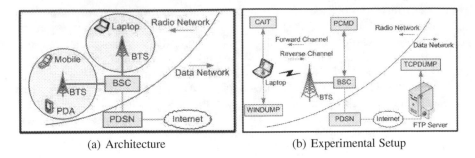

(a) Architecture   (b) Experimental Setup

Fig. 1. CDMA2000 Network Architecture and Experimental Setup

The main element of the radio network is the Base Station Controller (BSC), which is responsible for maintaining the *radio session* with the mobile device. The BSC controls hundreds of Base Transceiver Stations (BTS), which are essentially the air interfaces to the mobile devices, through a low-latency back-haul network. The importance of the BSC arises from the fact that it hosts two critical components which can directly affect higher layer performance: a) the *wireless scheduler* that dynamically controls the wireless channel rate assigned to the mobile device, and b) the *Radio Link Protocol* (RLP) that is responsible for fast MAC layer error recovery through the re-transmission of radio frames to recover from losses either over the low-latency back-haul connecting the BSC to the BTS or the wireless channel.

The function of the wireless scheduler is to assign a wireless channel rate (from up to *six* discrete rates) to a mobile device *on-the-fly*. This objective is primarily achieved by controlling the CDMA code length and channel power. Since higher rates require more power and resource reservation, the decision on when to allocate higher rates and to which user, must be made judiciously. In practice, the scheduler's decision could be influenced by three factors, which we investigate in detail in Section 4:

1. The queue length at the BSC (each user is assigned a separate buffer)
2. The channel conditions experienced by the mobile device, which is defined as the ratio between the received pilot signal strength E_c and the ambient noise and interference I_0, *i.e.*, E_c/I_0
3. The number of active users in the same sector (or sector load)

3   Experiments and Data Sets

Our primary focus is on the *downlink*. We performed end-to-end experiments which involved data transfer via either UDP or TCP SACK from a RedHat Linux server on the Internet to one or more laptops running Windows XP that were connected to the cellular data network via CDMA2000 1xRTT air-cards. A typical experimental setup is shown in Fig. 1(b) to illustrate the data path[1], as well as measurement points.

---

[1] The end-to-end path on average had a: a) propagation delay of 450-550ms, b) 25-35KB bottleneck buffer at BSC, c) 70KB-120KB wireless channel rate, and d) packet loss rate of 0.9%.

The experiments can be categorized into two classes. The first class consisted of sending UDP traffic to characterize the wireless scheduler. UDP was chosen to remove any transport layer feedback so that the wireless scheduler could be characterized in isolation. The second class comprised of downloading files via TCP in order to characterize long term TCP behavior, as well as its dependency on RF factors. These experiments were conducted under different TCP-specific and wireless configurations to evaluate their relative impact and obtain a better understanding of the system.

Each experiment, under every configuration, was run 10 times at various times during the day to obtain a reasonably large set of samples for statistical characterization. All plots include error bars denoting the 90% confidence interval around the mean. For TCP downloads, we used a single file size of 5MB since we are interested in long-term TCP behavior. The typical duration of both TCP and UDP flows was 15-20 minutes.

For each experiment, we collected data from the higher layer protocols through standard UDP/TCP logs at the client (windump) and server (tcpdump), as well as RF layer information. The RF statistics were collected from two observation points. Messages related to instantaneous channel quality, frame errors, re-transmissions and the assigned wireless channel rate were collected at the laptops using an air-interface analysis tool called CAIT [8]. These messages were aggregated to generate a time-series tracking the value of the above RF variables at a time-granularity of 1 second. The second source of measurement was the *Per Call Measurement Data* (PCMD) obtained from the BSC. PCMD contains details for all voice and data calls[2], such as source/destination phone numbers, cell/sector ID's, call duration, number of transmitted bytes, and call success/failure information. We used the PCMD logs to infer the number of active sessions in a sector (*i.e.,* sector load).

4   Wireless Scheduler Characterization

In this section we present an empirical evaluation of the factors that affect the behavior of the wireless scheduler. The exact implementation details of a commercial 1xRTT wireless scheduler is proprietary and hence we have to infer its behavior indirectly.

As mentioned in Section 2, channel rates assigned by the wireless scheduler can be influenced by the user's queue length, channel conditions and sector load. To understand the extent to which each of these factors affects the scheduler's decisions, we performed numerous UDP experiments with constant bit-rate (CBR) and 'on-off' traffic sources.

We begin by examining the impact of the application data rate on the wireless scheduler. Figure 2 plots the average throughput of a connection as a function of the data sending rate of a UDP CBR traffic source. The figure indicates that the achieved throughput tracks the sending rate up to a rate of 50 kbps after which it *decreases* sharply.

We next study *how* the assigned channel rate tracks the source's data rate and empirically show that the wireless scheduler assigns channel rates by tracking the user's *buffer backlog* over *short timescales*. Figure 4 plots the source's data rate and the assigned channel rate time-series (measured by CAIT) when the source utilizes an 'on-off' traffic pattern. The two figures show the data sending rate for two different on-rates (38.4kbps

---

[2] A single sector, covered by a BTS, typically had 8-9 (and a maximum of 30) active calls.

and 76.8kbps) and the same duty cycle (on for 1-second and off for 5-seconds). For the wireless scheduler to assign the correct channel rates for every 1-second burst of data transmitted by the sender, it must inspect the buffer backlog (or some function of it) at least once every second. This implies that the scheduler's decisions are very sensitive to the data sending rate determined by the transport and application layers.

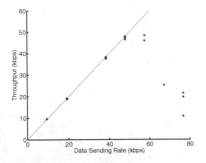

Variable	NMI Value
Buffer Backlog	0.57
Channel Quality (E_c/I_0)	0.15
Sector Load	0.11

Fig. 2. Throughput for UDP CBR experiments   **Fig. 3.** Factors affecting channel rate

Even though Fig. 4 indicates the presence of a strong correlation between the scheduler and the buffer backlog, visual identification is not always reliable or more importantly quantifiable, which is necessary for comparison. Consequently, we have developed an Information-Theoretic methodology based on a metric that we refer to as the *Normalized Mutual Information* to quantify the correlation between two time-series. We explain the main idea briefly. The Normalized Mutual Information (NMI) measures the amount of information [4] a time-series X can provide about another time-series Y (taking into consideration time-shifts between the two sequences). NMI lies between 0 and 1 and the larger the peak value, the more the two sequences are dependent on each other. NMI is basically a time-shifted correlation measure that can capture both linear and non-linear dependencies between time-series. For completeness, NMI is defined as:

$$I_N(X;Y;d) = (H(X) + H(Y) - H(X, Y_d))/H(X) \tag{1}$$

where $H(X)$ denotes the entropy of X, and $H(X, Y_d)$ denotes the joint entropy of X and a version of sequence Y that is time delayed by d time units[3].

Armed with this tool, we evaluated the impact of both the *channel conditions* and the *sector load* on the rate decisions made by the wireless scheduler, over short timescales, when sending UDP CBR traffic. We also computed the NMI between the buffer backlog and the channel assigned rate time-series for the 'on-off' traffic source. Figure 3 presents the peak NMI values indicating the *relative* impact that all three factors (averaged across all experiments) have on the channel assigned rates.

Quite surprisingly, as evident from the table, we found that the *short-term* channel quality and sector load have a very limited impact on the scheduler. We believe this

---

[3] When computing the peak NMI we consider all possible delay shifts d.

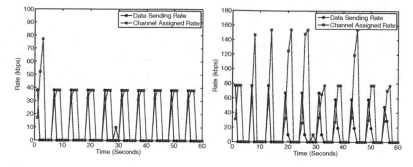

Fig. 4. Behavior of the wireless scheduler over short timescales. On-off 1s-5s at 38.4kbps (Left). On-off 1s-5s at 76.8 kbps (Right).

can be attributed to two reasons: a) fast power control[4] deployed in CDMA networks combats short-term channel fluctuations thus eliminating its impact on rate assignment, and b) the sector load varies too slowly to have a significant *short-term* impact on the scheduler's behavior. In Section 5 we show that these two factors affect *long-term* average scheduler behavior. The table also supports our initial observation that the *buffer backlog* is the most dominant factor influencing the scheduler's decisions.

5   TCP in CDMA2000 Networks

This section is devoted to results related to TCP behavior in experiments that were conducted under four different configurations on a commercial CDMA2000 network, where the following parameters were varied, namely: a) TCP's advertised receiver window size (ARWND), b) TCP's time-stamp option, c) number of active users in the sector, and d) user mobility (*i.e.*, speed) and location (*i.e.*, channel quality). The first two configurations are specific to TCP, while the latter two are more characteristic of wireless users. Clearly, there are several other variations (e.g. the TCP congestion mechanism) that are possible, however, we believe that the configurations we focused on are the most fundamental ones. Specifically, almost all operating systems have support for modifying TCP's ARWND and time-stamp options. Similarly, mobility and user location are the main characteristics of *any* wireless network.

Before discussing the experimental results, it is worthwhile making a general observation regarding our results. In almost all configurations, we found that the amount of variation in attained throughput is at least 10% and at times higher than 100%, even across consecutive runs of the same experiment, depending on location and time. Our measurements indicate that these variations in capacity were not caused by the channel quality or sector load. Instead, we believe they may be due to dynamic cell-dimensioning[5] performed by the network operator which is the focus of our future work.

---

[4] BTS boosts the transmitted signal's strength to increase the signal-to-noise ratio at the receiver.
[5] Adapting a BTS's maximum transmission power allows it to vary its coverage area.

5.1   TCP Window Size and Time-Stamp Option

The first two configurations we study involve TCP's behavior as a function of ARWND, both when the time-stamp option [10] was disabled/enabled. The window size was varied from 8 KB to 128 KB to control the number of packets in-flight in the network and hence the bottleneck queue size (equivalently queuing delay). This allowed us to subject the wireless scheduler to different queuing loads. Also, setting small advertised receiver windows (particularly ones that are smaller than the bottleneck buffer at the BSC) allowed us to emulate environments where wireless losses are more prevalent (*i.e.*, no congestion losses, due to buffer overflow, could ever occur). The time-stamp option, on the other hand, primarily allows the TCP sender to obtain more accurate RTT/RTO estimates which aid in detecting and avoiding spurious re-transmissions. Enabling/disabling this option allowed us to evaluate its impact on a connection's performance.

We begin by evaluating the impact of RF factors on a particular aspect of TCP, the round trip time (RTT). The NMI metric is used to quantify the relative impact of the RLP re-transmission rate, wireless channel rate, as well as the buffer occupancy (approximated by the number of un-acknowledged packets in-flight). Table 1 presents the peak NMI values between the RTT time-series and each of the three factors for a few different ARWND values. The table clearly indicates that for small and medium window sizes (8 KB and 16 KB) the wireless channel rate has the strongest influence on RTT, while at large window sizes (64 KB), buffer occupancy is high, and hence queuing delay becomes the dominant component in RTT.

These observations have several implications: a) RLP re-transmissions do not add significant latency, and b) in the absence of queuing, the channel rate determines RTT. Since RTT directly impacts throughput, we can expect TCP's throughput to be highly dependent on the assigned channel rates. This is indeed true as shown in Table 2 which presents the amount of information (NMI) that the channel rate and RLP re-transmission rate have about TCP's throughput. We showed previously in Section 4 that the channel rate is influenced by the transport data rate which implies that there is a strong coupling between the wireless scheduler and TCP. More importantly, it indicates that the rate variations in the wireless channel are *not* completely random, as is commonly assumed in models [1,3]. Instead, the channel's rate is *highly* correlated with TCP's state and must be taken into consideration in future models, and c) since the two controllers (TCP and wireless scheduler) take each other as input, this can lead to oscillations resulting in highly variable RTT, causing spurious re-transmissions.

We study the latter two issues in more detail. Figure 5 (Left) plots the cumulative TCP throughput as a function of ARWND. As a general observation, we note that TCP throughput increases as ARWND is increased, which is to be expected. First, let's consider the coupling between the scheduler and TCP. To highlight how this is different from an arbitrary random channel with some mean rate, we plot throughput obtained from an *ns*-2 simulation that has the same parameters as the active experiments except that the channel rate was set to the *average* assigned channel rate inferred from the CAIT logs. One can clearly see that the simulation predicts a far higher throughput than that obtained from the experiments (the lowest curve being the one with the time-stamps option disabled). Quantifying the exact relative impact of random channel variations (considering the timescale and magnitude of the variations) and coupling between scheduler and TCP (considering sensitivity of assigned channel rate to buffer backlog and vice versa) is part of our future work.

The second aspect we mentioned was that oscillations produced by such a coupling could result in highly variable RTT, causing spurious re-transmissions, including spurious timeouts [6]. To test this hypothesis, we ran experiments where the time-stamp option was enabled, and thus RTT variability is more accurately captured. Indeed, with time-stamps, the attained throughput is much higher as shown in Fig. 5 (Left).

Finally, we briefly look at packet losses as a function of window size. Figure 5 (Right) shows that packet losses increase with ARWND. When the window size is less than the bottleneck buffer at the BSC (25-35KB), all losses are due to the wireless channel. For window sizes larger than 35KB, *congestion* becomes the dominant cause of packet loss as the packet loss rate curve flattens out.

Table 1. Impact of various factors on RTT

(ARWND) Advertised Receiver Window Size	Peak NMI Values		
	Channel Rate	RLP	Packets in Flight
8 Kbytes	0.18	0.11	0.05
16 Kbytes	0.20	0.04	0.038
64 Kbytes	0.08	0.03	0.41

Table 2. Impact of RF factors on instantaneous TCP throughput

(ARWND) Advertised Receiver Window Size	Peak NMI Values	
	Channel Rate	RLP
8 Kbytes	0.16	0.06
64 Kbytes	0.26	0.03

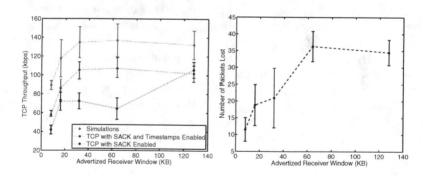

Fig. 5. TCP throughput (Left) and packet loss (Right) as a function of ARWND

5.2 Sector Load, User Mobility and User Location

Our final two configurations incorporate characteristic wireless behavior. For these experiments TCP's ARWND was set to the default (64 KB) with no time-stamp option.

We first varied the number of active TCP sessions (*i.e.,* data calls) within a single sector to study how TCP throughput changes with sector load, as well as evaluate the

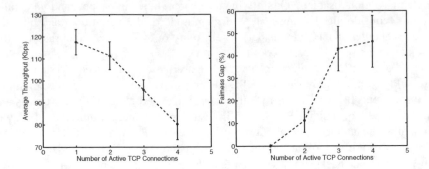

Fig. 6. Impact of sector load: throughput (Left) and fairness gap (Right)

wireless scheduler's fairness. Towards this end, we simultaneously downloaded files from up to 4 co-located laptops. The experiments were conducted during off-peak hours to ensure that the only users in the sector were the experiment laptops. In Fig. 6 we plot the cumulative TCP throughput (Left), as well as the Fairness Gap (Right) as a function of the number of active users. A perfectly fair scheduler would result in a Fairness Gap of 0. The larger the gap, the more unfair the scheduler. For any given set of throughput values $(y_1, y_2, ..., y_n)$, the Fairness Gap is defined as:

$$f_{\text{gap}}(y_1, y_2, ..., y_n) = \frac{\max(y_1, y_2, ..., y_n) - \min(y_1, y_2, ..., y_n)}{\min(y_1, y_2, ..., y_n)} \tag{2}$$

As expected, the average throughput achieved per user decreases as the number of active connections increases. However, we note that the fairness of the scheduler degrades with the number of active connections, as reflected by a larger Fairness Gap. Indeed, manual inspection of our experiments indicate that the throughput achieved by concurrent connections can be highly disparate with typically one user dominating.

The final configuration involved evaluating the impact of user mobility and location on the connection's performance. The mobility experiments were conducted on a 30-mile stretch of highway (RT 101) between San Francisco and Palo Alto, during non-peak hours (*i.e.*, at night). Connections lasted 10-15 minutes which is the time it takes

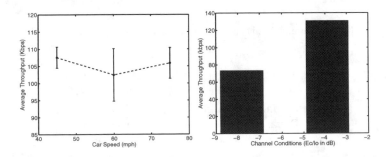

Fig. 7. Impact of user mobility (Left) and user location (Right)

to download a 5MB file. A BTS's coverage area is 2-3 miles causing a 10-15 minute connection to experience at least two hard hand-offs[6], assuming the car's speed is 45mph. Due to mobility, specifically path loss and shadowing, the mobile device experiences large variations in channel quality causing E_c/I_0 to fluctuate between 0dB (best possible channel) and -16dB (very poor channel). Figure 7 (Left) shows the achieved TCP throughput for three different average speeds of 45, 60 and 75 mph, respectively. Surprisingly, user speed had little impact on TCP throughput indicating that the cellular network is well engineered for fast hand-offs. We note that mobility is a major concern in 802.11 networks which are not *a priori* designed to handle fast transitions.

The last set of experiments were conducted to investigate the impact of *average (long-term)* channel conditions. In Section 4 we showed that the *short-term* scheduler behavior was not affected by *instantaneous* variations in channel conditions. However, it is unclear whether this observation carries over to *longer timescales*. To investigate this, we performed two sets of experiments, where the laptop was placed in locations with either *consistently good* or *bad* channels. The average throughput for each location is plotted in Fig. 7 (Right)[7]. One can clearly see that the throughput in locations with better channel conditions (*i.e.*, higher E_c/I_0) is much higher. This indicates that the long-term scheduler behavior is indeed affected by *average* channel conditions and not overcome by power control.

6   Conclusions

We conducted a detailed cross-layer measurement study to evaluate TCP behavior over CDMA2000 networks. The study was conducted under various configurations that involved simple variations of TCP, as well as, changing user mobility and sector load. By conducting measurements across all three (*i.e.*, transport, MAC and RF) layers, we were able to evaluate the system comprehensively. Our main findings were:

1. The RLP layer allows fast error recovery that almost eliminates packet loss observed at the transport layer, while having a minimal impact on TCP's RTT.
2. The wireless scheduler in CDMA2000 networks is unaffected by channel conditions or sector load over short timescales. Instead, the decisions are highly dependent on the transport data rate. However, the long-term scheduler rate allocation is indeed affected by *average* channel conditions and sector load. Furthermore, increasing sector load deteriorates the fairness of the scheduler.
3. The wireless scheduler and TCP are strongly coupled which can result in highly variable RTT. Apart from modeling implications since the rate variations are not completely random, it motivates the need for robust RTT estimation to prevent spurious re-transmissions.
4. Mobility is well supported in the CDMA2000 network and hence had no major impact on TCP throughput.

---

[6] A hard hand-off occurs when the BSC completely switches data from one BTS to another.
[7] The RTT, general path characteristics and variation in E_c/I_0, for both locations, were very similar.

References

1. E. Altman, C. Barakat, and V. M. R. Ramos. Analysis of AIMD protocols over paths with variable delay. In *Proc. IEEE INFOCOM*, Hong Kong, March 2004.
2. P. Benko, G. Malicsko, and A. Veres. A Large-scale Passive Analysis of End-to-End TCP Performance over GPRS. In *Proc. IEEE INFOCOM*, Hong Kong, 2004.
3. M. C. Chan and R. Ramjee. TCP/IP Performance over 3G Wireless Links with Rate and Delay Variation. In *Proc. ACM MOBICOM*, pages 71–82, 2002.
4. T. M. Cover and J. A. Thomas. *Elements of Information Theory*. Wiley-Interscience, 1991.
5. Y. Lee. Measured TCP Performance in CDMA 1xEV-DO Network. In *PAM*, Adelaide, Australia, 2006.
6. R. Ludwig and R. H. Katz. The Eifel Algorithm: Making TCP Robust Against Spurious Retransmissions. *SIGCOMM*, 30(1):30–36, 2000.
7. K. Mattar, A. Sridharan, H. Zang, I. Matta, and A. Bestavros. TCP Over CDMA2000 Networks : A Cross-Layer Measurement Study. Available at http://research.sprintlabs.com, Sprint ATL, October 2006.
8. Qualcomm. CDMA Air Interface Tester. www.cdmatech.com/products/cait.jsp
9. J. Ridoux, A. Nucci, and D. Veitch. Seeing the difference in IP Traffic: Wireless versus Wireline. In *Proc. IEEE INFOCOM*, Barcelona, Spain, 2006.
10. D. Borman V. Jacobson, R. Braden. RFC 1323: TCP Extensions for High Performance.

A Measurement Study of Scheduler-Based Attacks in 3G Wireless Networks

Soshant Bali[1], Sridhar Machiraju[2], Hui Zang[2], and Victor Frost[1]

[1] University of Kansas
{sbali,frost}@ittc.ku.edu
[2] Sprint ATL
{Machiraju,Hui.Zang}@sprint.com

Abstract. Though high-speed (3G) wide-area wireless networks have been rapidly proliferating, little is known about the robustness and security properties of these networks. In this paper, we make initial steps towards understanding these properties by studying Proportional Fair (PF), the scheduling algorithm used on the downlinks of these networks. We find that the fairness-ensuring mechanism of PF can be easily corrupted by a malicious user to monopolize the wireless channel thereby starving other users. Using extensive experiments on commercial and laboratory-based CDMA networks, we demonstrate this vulnerability and quantify the resulting performance impact. We find that delay jitter can be increased by up to 1 second and TCP throughput can be reduced by as much as $25 - 30\%$ by a single malicious user. Based on our results, we argue for the need to use a more robust scheduling algorithm and outline one such algorithm.

1 Introduction

Today, the mobile Internet is one of the fastest-growing segments of the Internet. One of the main reasons for this is the rapid adoption of high-speed (3G) wide-area wireless networks. The two main 3G standards are Evolution Data Optimized (EV-DO) [1] and High-Speed Downlink Packet Access (HSDPA). The increasing use of these networks to access the Internet makes it important that these are well-engineered, robust and secure. Much work has been done on designing these networks [3,7,5] and the algorithms used in them, especially, the wireless scheduling algorithms [4,11,13]. However, most prior work has focused on improving system performance assuming cooperative scenarios without malicious users.

In this paper, we make some initial steps towards understanding the important issue of 3G robustness from the security viewpoint. We study the scheduling algorithm since it plays a vital role in deciding system performance and user experience. Some prior work [9] have studied vulnerabilities associated with FIFO schedulers that are common on wired IP networks. Since most 3G networks use the Proportional Fair (PF) algorithm [7] for *downlink* scheduling, we focus on it in this paper. PF has been widely deployed because it is simple and increases system throughput by being channel-aware, i.e., it schedules data transmission

S. Uhlig, K. Papagiannaki, and O. Bonaventure (Eds.): PAM 2007, LNCS 4427, pp. 105–114, 2007.

to users with good wireless conditions over those who are experiencing *fading* (bad channel conditions). Under general conditions, PF maximizes the product of throughputs received by all users [13]. Moreover, when all users have identical and independent fading characteristics, PF is known to be fair in the long term.

The main finding of our study is that the fairness-ensuring mechanism of PF can easily be corrupted by a malicious user to monopolize the wireless channel thereby starving other users. Such scheduler-based attacks are possible because PF does not distinguish between users with outstanding data and those without. Using extensive measurements on a commercial CDMA network and on a similar laboratory setup, we show that the performance degradation due to such attacks is severe - they can increase "jitter" by up to 1 second and cause frequent spurious TCP timeouts. We also show that the latter can increase flow completion times and decrease TCP goodput by up to 30%. Our findings are important not only because we tackle mechanisms used in 3G networks and possibly, other future wireless networks, but also because our work (re-)emphasizes the need to consider security while designing network algorithms.

This paper is organized as follows. Section 2 provides an overview of the PF scheduling algorithm and describe how it can be attacked. In Section 3, we conduct initial experiments on a commercial network and motivate the need to move to a more controlled experimental environment. In Section 4, we use experiments in our laboratory to quantify the impact of PF-induced attacks on UDP and TCP-based applications. We also discuss other important issues including a possible replacement for PF. We conclude and outline future work in Section 5.

2   The PF Algorithm and Starvation

As with any managed wireless network, access to the wireless channel in 3G networks is controlled by Base Stations (BSs) to which mobile devices or Access Terminals (ATs) are associated. Our focus is on Proportional Fair (PF) - the scheduling algorithm [13] used to schedule transmissions on the downlink in most 3G networks. In these networks, downlink transmission is slotted. For example, in CDMA-based EV-DO networks, slot size is 1.67ms. BSs have per-AT queues and employ PF to determine the AT to transmit to, in a time slot.

The inputs to PF are the current channel conditions reported on a per-slot basis by each AT. Specifically, each AT uses its current Signal-to-Noise Ratio (SNR) to determine the required coding rate and modulation type and hence, the *achievable rate* of downlink transmission. In the EV-DO system, there are 10 unique achievable data rates (in Kilobits per second) - 0, 38.4, 76.8, 153.6, 307.2, 614.4, 921.6, 1228.8, 1843.2 and 2457.6. Assume that there are n ATs in the system. Denote the achievable data rate reported by AT i in time slot t to be R_t^i ($i = 1 \ldots n$). For each AT i, the scheduler also maintains A_t^i, an exponentially-weighted average rate that user i has achieved, i.e.,

$$A_t^i = \begin{cases} A_{t-1}^i(1 - \alpha) + \alpha R_t^i & \text{if slot allocated} \\ A_{t-1}^i(1 - \alpha) & \text{otherwise} \end{cases}$$

Slot t is allocated to the AT with the highest ratio $\frac{R_t^i}{A_{t-1}^i}$. α is usually around 0.001 [7] (we verified this using measurements [2]). Thus, an AT will be scheduled less often when it experiences fading and more often when it does not. Under general conditions, PF maximizes the product of per-AT throughputs [13]. Moreover, if all ATs have identical and independent fading characteristics, all ATs are allocated equal number of slots in the long term. If different ATs experience non-identical wireless conditions, unequal slot allocation may result [4] in the long term. Long-term fairness is not our focus here. We show how malicious ATs can cause short-term unfairness that is severe enough to degrade application performance.

The basic observation behind our work is that a malicious AT can influence the value of its $\frac{R_t}{A_{t-1}}$ ratio thereby affecting the slot allocation process. An AT can do this simply by receiving data in an *on-off* manner. To see why, consider an AT that receives no data for several slots. Its A_t would slowly reduce and approach zero. After several inactive slots, when a new packet destined for that AT arrives at the base station, that AT has a low value of A_t and is likely to get allocated the slot because its ratio is very high. This AT keeps getting allocated slots until its A_t increases enough. During this period, all other ATs are starved. A few Prior work [8] has observed excessive delays with PF. To our knowledge, no prior work has considered on-off traffic or explored how malicious users can exploit it.

Starvation due to on-off behavior occurs because PF reduces A_t during the off periods. This implies that PF "compensates" for slots that are not allocated even when an AT has no data to receive! In the rest of the paper, we study how a malicious AT (and a cooperating data source to that AT) can exploit this vulnerability, namely, the inability of PF to distinguish between an AT to which no data needs to be sent and one that is not allocated a slot.

3   Experimental Setup

We conduct our initial experiments on a commercial EV-DO network in USA. Our ATs are IBM T42 Thinkpad laptops running Windows XP equipped with commercially-available PCMCIA EV-DO cards. The laptops have 2GHz processors and 1GB of main memory. All ATs connect to the same base station and sector. Data to the ATs is sourced from Linux PCs with 2.4GHz processors and 1GB of memory. All of these PCs are on the same subnet and about $10-11$ hops away from the ATs.

For our first experiment, we use two ATs - AT1 and AT2. AT1 receives a long-lived periodic UDP packet stream consisting of 1500-byte packets with an average rate of 600Kbps. AT2 is assigned the role of a malicious AT and hence, receives traffic in an on-off pattern from its sender. Specifically, it receives a burst of 250 packets of 1500 bytes every 6 seconds. We plot the "jitter" experienced by AT1 in Figure 1. Since our ATs are not time-synchronized with the senders, jitter is calculated as the excess one-way delays over the minimum delay. Well-defined increases in jitter are observed whenever a burst is sent to AT2. In contrast, a base station employing fair queueing would cause almost no increase in "jitter" as

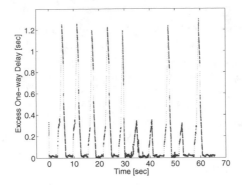

Fig. 1. "Jitter" caused by a malicious AT in a commercial EV-DO network

long as the wireless link capacity is not exceeded. These results clearly show that AT1 experiences extraordinary increase in "jitter". We observe similar results with other parameter settings (results not shown). With all of them, however, the jitter increases vary from 300ms to 1 second. The variability is likely due to traffic to other ATs and queueing effects at other hops. Hence, to understand and quantify the attack scenarios better, we move to a more controlled laboratory setup that recreates actual network conditions while eliminating unknowns such as cross-traffic. We describe this setup now.

In our laboratory, we use commercially available equipments that are widely used in commercial networks to recreate the EV-DO network including the Base Station, the Radio Network Controller (RNC) and the Packet Data Serving Node (PDSN) (see [7]). The links between the Base Station, RNC and PDSN are 1Gbps Ethernet links. The Base Station serves 3 sectors of which we use one. Our ATs and senders are the same as before. We collect *tcpdump* [12] traces at the senders and ATs. Due to the peculiarities of PPP implementation on Windows XP, the timestamps of received packets are accurate only to 16ms. However, these inaccuracies are small enough to not affect our results. For TCP-based experiments, we use *tcptrace* [12] to analyze the sender-side traces. There are three main differences with a commercial network. First, we use lower power levels than commercial networks due to shorter distances and in the interest of our long-term health. Since our goal is not to characterize fading and PF's vulnerability does not depend on channel characteristics, this does not affect the validity of our results. Second, we can control the number of ATs connected to the base station. Third, the number of hops from the senders to the ATs is only 3. This eliminates the additional hops on normal Internet paths and queueing effects on those hops. We discuss the impact of this in Section 4.2. Moreover, this is realistic in networks that use split-TCP or TCP-proxy [14].

Our laboratory setup poses a few challenges; we describe two of the more important challenges now. First, even though we conduct our experiments in the laboratory, the wireless conditions varied significantly. Hence, we conducted up to 30 runs of each experiment (a particular parameter setting) to calculate a

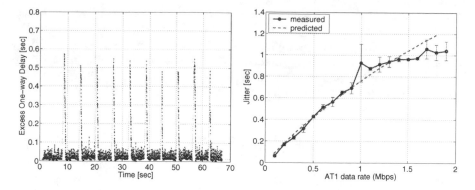

Fig. 2. (Left) Results of "jitter" experiment performed in the lab setup. We plot the excess of (unsynchronized) delays. (Right) The maximum amount of "jitter" - measured and predicted - that can be caused as a function of the data rate of the long-lived flow to AT1. As noted before, fair queueing would cause negligible "jitter" if channel capacity is not exceeded.

good estimate of the required performance metric with a small enough confidence interval. We also interleaved the runs of the different parameter settings used to plot a figure so that they all experienced the same wireless conditions on average. A second challenge is that ATs become disassociated with the base station after around 12 seconds of inactivity. Also, the initial few packets sent to an inactive AT encountered large delays due to channel setup and other control overhead. To prevent our ATs from becoming inactive, we use a low rate background data stream of negligible overhead.

4   Experimental Results

In this section, we use our laboratory setup to quantify the severity of PF-based attacks. We first focus on non-reactive UDP-based applications and then, on TCP-based applications. We conclude this section by discussing how common traffic patterns can also trigger PF-induced starvation and briefly discuss a preliminary replacement for PF.

4.1   UDP-Based Applications

In Figure 2, we plot the results of a laboratory experiment similar to that of Figure 1. We send a long-lived UDP flow of average rate 600Kbps to AT1 and bursts of 150 packets to AT2 every 6 seconds. The results mirror the behavior observed in the commercial network, namely, large "jitter" whenever a burst is sent. Notice the reduction in the variability of results due to the absence of other ATs and queueing at other hops.

Recall that the PF algorithm compares the ratios of all ATs to allocate slots. Intuitively, the "jitter" of AT1 depends on the value of A_t of AT1 just before a burst to AT2 starts. This can be analytically derived. Due to lack of space, we skip the derivation and only provide the final expression for the "jitter" J experienced by AT1 when both ATs experience unchanging wireless conditions and hence, have constant achievable data rates $R1_t = R1$ and $R2_t = R2$ in every time slot t. We also assume that $A1_T = \beta 1_T R1$ and $A2_T = \beta 2_T R2$ are the moving averages for AT1 and AT2 in time slot T, the last slot before a burst to AT2 starts. Then (the derivation can be found in [2]),

$$J = \left\lceil \frac{log(\frac{1}{1+\beta 1_T - \beta 2_T})}{log(1 - \alpha)} \right\rceil \tag{1}$$

In Figure 2 (Right), we plot the predicted value of "jitter" assuming $R1 = 1.8$Mbps and $\beta 2_T = 0$. We compare this with experiments in which we vary the rate of AT1's flow from 100Kbps to 2Mbps. For each experiment, we calculate the maximum "jitter" experienced by AT1 and plot them in Figure 2. Comparing the results of these experiments with $\beta 2_T = 0$ makes sense because the bursts are separated long enough that AT2's A_t is close to zero. It is clear that the experimental results closely follow the analytically predicted values. Also, the jitter experienced by AT1 increases almost linearly with the *entire* data rate to AT1. Thus, an AT1 with a single VoIP application of 100Kbps may experience only 100ms increase in "jitter" whereas additional concurrent web transfers by this VOIP user would cause larger "jitter". As another example, an AT receiving a medium-rate video streams of 0.6Mbps could experience a jitter increase of more than 0.5 seconds. This can cause severe degradation in video quality.

4.2   Effect on TCP Flows

We now show that TCP-based applications are also susceptible to PF-induced attacks. We start by conducting an experiment in which we replace the UDP flow to AT1 with a long-lived TCP transfer of 20MB. As before, we send an on-off UDP stream to AT2 in which every burst consists of 150 1500-byte packets once every 3 seconds - an average rate of 600Kbps. We analyze sender-side *tcpdump* [12] traces with *tcptrace* [10] and plot the TCP sequence number of the bytes transmitted versus time of the flow to AT1 in Figure 3 (Left). The SYN packet is marked at time 0. The black dots represent transmitted packets (x-value is time of transmission and y-value is the sequence number). We see periodic retransmissions (blobs of small Rs) every 3 seconds corresponding to each burst of the flow to AT2. This demonstrates how a malicious user can easily cause TCP timeouts to other users.

TCP timeouts in the above experiment could be caused due to one of two reasons. The first reason is that AT1 is starved long enough that its buffer overflows and some packets are dropped. The second reason is that the buffer is large enough but AT1's packets are delayed long enough that TCP experiences a *spurious timeout*. It turns out that per-AT buffers in EV-DO base stations are usually 80 to 100KB in size, which is larger than the default receiver window size of 64KB in Linux and Windows (other versions use 32KB and 16KB [6]). We also verified this using sender side *tcpdump* traces. Due to lack of space, we do not explore scenarios involving multiple flows to the same AT and timeouts caused by the resulting

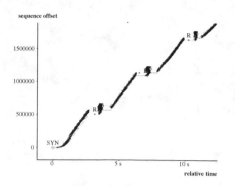

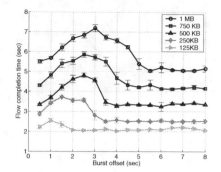

Fig. 3. Results of *tcptrace* analysis of AT1. Timeouts are caused whenever AT2 received a burst.

Fig. 4. Increase in flow completion time for short TCP flows. 95% confidence intervals are plotted.

packet losses. It might be argued that our laboratory setup causes more spurious timeouts because we have fewer hops than typical Internet paths. In fact, our setup reflects the common practice of wireless providers in using split-TCP or TCP proxies [14]. Moreover, as wireless speeds go up, delays are only going to decrease.

Short Flows. We now study the impact on TCP performance due to spurious timeouts caused by a malicious user. We first consider short TCP flows for which flow completion times are the suitable performance metric. We conduct experiments as before but replace the UDP flow to AT1 with TCP transfers ranging from 125KB to 1MB. Since short flows spend a significant fraction of time in slow start, A_t is likely to be small early on. Hence, the starvation duration is likely to depend on the offset of the burst from the start time of the TCP flow. To understand this better, we conduct experiments for various values of the burst offsets. For each offset and flow size, we run 30 experiments and plot the average flow completion time in Fig. 4. We make four observations. First, for a large enough offset, the burst has no impact because the TCP flow is already complete. Second, the probability of a timeout increases as the offset increases. This confirms our intuition that, during slow start, A_t of AT1 is smaller and hence, starvation duration is smaller. Maximum performance impact is realized when the offset is $2 - 3$ seconds. This is observed when we plot the average number of retransmissions too (figure not shown due to lack of space). Third, the inverted-U shape shows that the probability of a timeout decreases when the burst starts towards the end of the flow. Fourth, for downloads of 250KB and above, there is a $25 - 30\%$ increase in flow completion time. Note, however, that A_t depends on the *total* data rate to AT1. Hence, if AT1 receives other data flows simultaneously, its A_t would be larger and more timeouts may result.

Long Flows. Next, we study long-lived TCP flows for which the suitable performance measure is goodput. Consequently, we start a long TCP flow to AT1. Our malicious AT, AT2, receives on-off traffic in the form of periodic bursts. To understand how AT2 can achieve the maximum impact with minimal overhead,

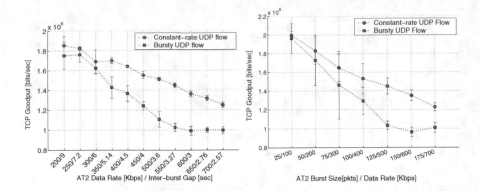

Fig. 5. Plots illustrating the reduction in TCP goodput as a function of the burst size (Left) and burst frequency (Right) of an on-off UDP flow. Note that the y-axis on the plots does not start at 0.

we conduct experiments with various burst sizes and frequencies. Since the average rate to AT2 changes based on the burst size and frequency, we cannot compare one experiment to another. Instead, we compare each experiment with an experiment in which AT2 receives a constant packet rate UDP stream of the same average rate. The TCP goodput achieved with such well-behaved traffic captures the effect of the additional load. Any further reduction in goodput that we observe with on-off UDP flows essentially captures the performance degradation due to unnecessary timeouts. We plot the average TCP goodput achieved in our experiments with on-off and well-behaved UDP flows to AT2 in Figure 5. In the Left plot, we vary the inter-burst gap for a burst size of 150 1500-byte packets. As expected, the slope of goodput with well-behaved UDP flows is almost linear with slope close to -1. The performance impact of malicious behavior is clearly shown with the maximum reduction in goodput when the inter-burst gap is around $3 - 3.5$ seconds. In this case, the goodput reduces by about 400Kbps - almost 30%. Larger gaps cause fewer timeouts and smaller gaps cause bursts to be sent before AT2's A_t has decayed to a small enough value. In the Right plot, we vary the burst size for a 3-second inter-burst gap. We find that bursts of $125 - 150$ packets cause the largest reduction in goodput of about $25 - 30\%$.

4.3   Discussion

Starvation-driven spurious timeouts can also be triggered accidentally by benign users with typical user behavior (also, see [8]). We illustrate such a scenario using an experiment in which AT2 periodically downloads a 500KB file via TCP (to model HTTP transfers). AT1 receives a long-lived TCP flow. We plot the round-trip times and TCP sequence numbers of AT1 in Fig. 6. We see large RTT increases corresponding to the slow start phase of AT2's periodic downloads. In the TCP sequence plot (right), we see that many of these cause timeouts. For this experiment, AT2 is at a spot where it had a smaller average achievable rate

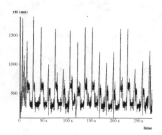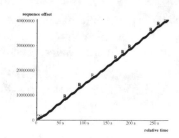

Fig. 6. Starvation caused by periodic short TCP flows to AT2 on a long-lived TCP flow to AT1. (Left) RTT over time. (Right) TCP sequence number over time.

Table 1. Performance improvement with our preliminary solution

Scenario	AT1 (TCP)	AT2	Perf. Metric	Improvement
I	long flow	periodic UDP burst	goodput	82.40%
II	short flow 1MB	periodic UDP burst	completion time	28.10%
III	short flow 500KB	periodic UDP burst	completion time	26.64%
IV	long flow	periodic TCP download 50KB	goodput	8.03%

(921.6Kbps) so that it can cause significant starvation to AT1 even when it is in the slow start phase. In future work, we intend to thoroughly explore the kinds of user behavior that can accidentally trigger starvation and to determine how frequently they occur in practice.

To rectify the vulnerability detailed in this paper, PF needs to be modified. A naive strategy is to stop updating A_t when an AT has no data to receive. This does not work because A_t needs to reflect the total number of active users *and* the recent channel conditions of that AT - both of which might change. For similar reasons, we cannot use the A_t values of ATs already receiving data either. Investigating a complete solution is out of the scope of this paper. Here we briefly discuss a promising candidate. In our proposed algorithm, we run two instances of PF - one of which assumes that all ATs always have data to receive. Actual slot allocation is governed by the second instance, which resets its A_t values with those of the first instance whenever an idle AT begins to receive data.

We explore the potential of the above solution using *ns-2* simulations of four scenarios corresponding to those explored previously and summarize the results in Table 1. In the simulation we use traces of achievable rate collected from the commercial EV-DO network. AT2 assumes the role of a malicious user in scenarios I, II, and III and of a benign user who accidentally causes AT1 to timeout in scenario IV. Since the timeout calculation in *ns-2* results in a larger value than Linux, we had to use an achievable data rate of 300Kbps for AT2 to simulate the accidental timeouts observed in Figure 6. In Table 1, we list the performance improvement, which is the decrease in completion times of a short flow or increase in TCP goodput of a long flow for AT1, of our proposed algorithm over PF in the four scenarios. We see that our solution is more robust to malicious behavior. In fact, we verified that it eliminates virtually all spurious timeouts that PF may introduce. A detailed investigation into alternative scheduling algorithms and their performance is in the longer version of this paper [2].

5   Conclusions and Future Work

We showed that the Proportional Fair (PF) algorithm, which is used in many 3G networks, can be easily corrupted by malicious users to starve users. Using extensive measurements, we showed that such starvation can occur in deployed systems and lead to severe performance degradation including jitter of more than a second and spurious TCP timeouts. The latter can reduce TCP goodput and increase flow completion times by about 30%. As future work, we intend to extend our study to scenarios with more than two ATs, where we expect the results to follow our analytical predictions as in this paper. We briefly explored a promising algorithm to overcome PF's vulnerability. We intend to explore this more thoroughly in the future.

References

1. Telecommunications Industry Association. CDMA 2000: High Rate Packet Data Air Interface Specification (TIA-856-A), 2004.
2. S. Bali, S. Machiraju, H. Zang, and V. Frost. On the Performance Implications of Proportional Fairness (PF) in 3G Wireless Networks. Technical Report RR06-ATL-040624, Sprint ATL, 2006.
3. P. Bender, P. Black, M. Grob, R. Padovani, N. Sindhushayana, and A. Viterbi. CDMA/HDR: A Bandwidth-efficient High-speed Wireless Data Service for Nomadic Users. *IEEE Communications Magazine*, 38:70–77, July 2000.
4. S. Borst. User-level Performance of Channel-aware Scheduling Algorithms in Wireless Data Networks. In *Proc. of IEEE INFOCOM*, 2003.
5. Mun Choon Chan and Ramachandran Ramjee. Improving TCP/IP Performance over Third Generation Wireless Networks. In *Proc. of IEEE INFOCOM*, 2004.
6. Carl Harris. Windows 2000 TCP Performance Tuning Tips
   http://rdweb.cns.vt.edu/public/notes/win2k-tcpip.htm.
7. A. Jalali, R. Padovani, and R. Pankaj. Data Throughput of CDMA-HDR: A High Efficiency-high Data Rate Personal Communication Wireless System. *Proc. of IEEE Vehicular Technology Conference*, 3:1854–1858, May 2000.
8. T. Klein, K. Leung, and H. Zheng. Enhanced Scheduling Algorithms for Improved TCP Performance in Wireless IP Networks. In *Proc. of GLOBECOM*, 2004.
9. A. Kuzmanovic and E.W. Knightly. Low-rate TCP-targeted Denial of Service Attacks: The Shrew vs. The mice and Elephants. In *Proc. of SIGCOMM*, 2003.
10. Shawn Ostermann. tcptrace, http://jarok.cs.ohiou.edu/software/tcptrace.
11. S. Shakkottai and A. Stolyar. Scheduling Algorithms for a Mixture of Real-time and Non-real-time Data in HDR. In *Proc. of ITC-17*, September 2001.
12. tcpdump. http://www.tcpdump.org.
13. P. Viswanath, D. Tse, , and R. Laroia. Opportunistic Beamforming using Dumb Antennas. *IEEE Transactions on Information Theory*, 48:1277–1294, June 2002.
14. W. Wei, C. Zhang, H. Zang, J. Kurose, and D. Towsley. Inference and Evaluation of Split-Connection Approaches in Cellular Data Networks. In *Proc. of PAM*, 2006.

Understanding Urban Interactions from Bluetooth Phone Contact Traces

Anirudh Natarajan, Mehul Motani, and Vikram Srinivasan

Electrical & Computer Engineering, National University of Singapore

Abstract. The increasing sophistication of mobile devices has enabled several mobile social software applications, which are based on opportunistic exchange of data amongst devices in proximity of each other. Examples include Delay Tolerant Networking (DTN) and PeopleNet. In this context, understanding user interactions is essential to designing algorithms which are efficient and enhance the user experience. In our experiment, users were handed Bluetooth enabled phones and asked to carry them all the time to log information about other devices in their proximity. Data was logged over several months, with over 350,000 contacts logged and over 10,000 unique devices discovered in this period.[1] This paper analyzes this data by charactering the distributions of metrics such as contact time and inter-pair-contact time, and introducing several other important metrics useful for understanding user interactions. We find that most metrics follow a power law, except for inter-pair-contact time. We also look for patterns in user interactions, with the hope that these can be exploited for better algorithm design.

1 Introduction

The increase in the capabilities of mobile communication devices has led to a plethora of interesting new applications. One category of these applications are Delay Tolerant Networking (DTN) applications such as Haggle [3], PeopleNet [8] and Serendipity [4]. The performance of these DTN application critically depends on patterns of user interactions. Previous work such as [3], [4] and [6] have attempted to study and characterize these interactions. However these studies were limited in scope and scale. Our experiment was done in the urban setting of Singapore where mobile phone penetration rates are exremely high. Reports from the Infocomm Development Authority of Singapore (IDA) [10], quote it to be 95.5%, more than a year ago. While we cannot exacty estimate the Bluetooth phone penetration rate, from our data we were able to find one new Bluetooth device roughly every 10-15 minutes.

Select people, whom we call probes, were given Bluetooth mobile phones and asked to carry the phones all the time to log information about other devices in their proximity. The phones handed out would perform Bluetooth device discoveries every 30 seconds and log the Bluetooth name of the external device logged,

---

[1] The dataset will be made available to the research community.

S. Uhlig, K. Papagiannaki, and O. Bonaventure (Eds.): PAM 2007, LNCS 4427, pp. 115–124, 2007.

the MAC address of the device and the time of the rendezvous. Data was logged over four months, with over 350, 000 logs and over 10, 000 unique devices discovered in this period. Compared to other studies employing a similar approach, our study is of a much larger scale. We discuss these differences in detail in Section 2. Section 3 lays out the details regarding how we went about collecting the data and the challenges we faced.

In Section 4 we characterize user interactions via the usual distributions of metrics of contact time, inter-pair-contact time (between specific pairs of devices) and inter-contact time (between any two devices). We also introduce new metrics, that shed light on the way people cluster based on the notion of a meeting. A meeting is an event that occurs when a user is in contact with at least one other device for some minimum duration.

Finally, in Section 5, we look at patterns in user interactions, with the premise that predictability can lead to the design of better opportunistic algorithms for DTN-like aplications. We looked at the time series of the number of contacts over different time scales and also looked for the occurence of common contacts over these time scales.

2   Related Work

There have been two kinds of trace based studies in the recent past. The first collects and analyzes traces collected from WiFi device association patterns with access points [1, 2, 5, 7, 11]. In the second approach, which is adopted by our study, users were handed Bluetooth based devices and interactions encountered by these users were studied.

One of the largest and first studies stems from the Serendipity [4] project at MIT. In their study, 100 users at MIT were handed Nokia 6600 phones and asked to carry them around for around 9 months. Apart from logging usage patterns

Table 1. A comparison of the different studies of peer-to-peer contact pattern traces between Bluetooth devices in terms of scope, duration, and amount of data collected

	Our Study	Haggle/ Intel	Haggle/ Cambridge	Haggle/ Infocom	Toronto	Serendipity
Device	Phone	iMote	iMote	iMote	PDA	Phone
Devices participating	12 (3 static, 9 mobile)	8	12	41	23	100
Duration	4 Months	3 days	5 days	3 days	16 days	9 months
Location	City-wide	Intel Campus	Campus	Conference	Campus	City-wide
Granularity	30 secs.	120 secs.	120 secs.	120 secs.	120 secs.	300 secs.
Logs	362,599	2,264	6,736	28,250	2,802	NA
Unique devices discovered	10,673	92	159	197	N/A	2798

of the phone, such as how often different applications on the phone were used, they also studied interactions with other test users by scanning for Bluetooth devices every 5 minutes. The University of Toronto study [6], and the Haggle studies [3], gave out between 8 and 41 Bluetooth enabled devices over a course of a few days and analyzed contact and inter contact times.

Table 1 shows the difference between our studies and the others. It is clear that the size and scope of our data is much larger to the ones that have been obtained in the other studies. We discovered more than 10, 000 unique devices and had over 350, 000 logs. This is orders of magnitude larger than data collected by any of the other studies. The granularity at which we chose to do our Bluetooth discovery is at an interval of 30 seconds. Other studies use a device discovery period of at least 120 seconds. Note that one could not go much lower in granularity as the Bluetooth discovery process takes roughly 10-20 seconds to complete. Moreover, we collected data over 4 months (with 296 man days of logging) which is larger than the Haggle and Toronto studies. Although [4] had a much larger number of volunteers, who logged data over a much larger duration, we notice that the number of unique devices discovered is much smaller. We believe that this is due to the fact that at the time the study was done, Bluetooth penetration rates were much lower. Moreover, device discovery was performed only every 5 minutes. When we looked at their data we found it difficult to mine statistics such as contact duration and inter-contact times accurately.

Moreover, the Haggle study etc., analyses only contact time and inter contact time distributions. However, with the constant increase in the number of applications based on opportunistically exploiting the proximity of devices it is clear we need to analyze several additional parameters.

For example in PeopleNet [8], the goal is not to transfer information from a certain source to destination. PeopleNet is a large distributed geographic database. Queries hop around from one mobile device to another in search of information that resides on these devices. In such a scenario, it is clear that the performance of the system is determined by how people aggregate rather than precise contact patterns between pairs of devices. Another example, stems from the notion of using sensors embedded in mobile phones to gather data from the environment. In [9], the authors propose an algorithm to aggregate data from different mobile devices which exploits the aggregation patterns of people. In these contexts, it is clear that one needs to understand aggregation dynamics. In our analysis of the data, we have designed metrics to understand these aggregation patterns.

3   Methodology

To allow us to get a wide variety of data we chose 12 probes [2]. Of these 3 were static and 9 were mobile. The static devices were customized, line powered, Bluetooth access points running on embedded Linux and these were placed in three of the busiest lecture theaters on National University of Singapore campus. The 9 mobile probes were chosen to get as diverse a sampling of various social behavior patterns.

---

[2] Due to certain logistical constraints we could not have more than 12 probes.

5 students on campus, 2 faculty members and 2 students who lived off campus carried mobile phones with the software that logged the Bluetooth device discoveries. After collecting the data we did realize that our choices did give us a varied set of behaviors. As expected, the 2 students living off campus logged the most contacts, logging around 170 distinct devices for every man day logged. Interestingly, the static probes discovered the least number of distinct devices per day. The maximum was 13.2 distinct devices per day. This clearly highlights the importance of mobility to increasing the potential for opportunistic data relay algorithms.

The reason phones were chosen instead of iMotes was that phones are personal devices that people already have a reason to carry around. This meant that users would remember to recharge the phones and always carry it with them over long durations (months). Further, mobile phones have more than 6MB of memory whereas iMotes have only 64KB.

Having narrowed down the choice, we picked Nokia 6600 and Panasonic X800 phones as they were the most reliable. In particular, HP PDA's and Sony Ercisson's consistently logged fewer devices than the former two devices under identical conditions.

The phones and the static devices conducted Bluetooth device discoveries every 30 seconds and logged the MAC addresses, the date and the time when the device was found. The static devices were programmed to upload their data to a central MySQL server once every day. The mobile probes had to transfer their data by activating a program on their computers that would then automatically transfer the data from the PC to the central server.

The main challenge faced in collecting the data was the finite battery life. Due to Bluetooth device discovery being an energy consuming process, phones would run out of power and the logging would stop. Often phones needed to be recharged every day in order to log continuously. Another source of error was human error. Despite our persistent attempts to remind the probes to keep the logging program switched on at all times, they had a tendency to switch it on in crowded areas which skewed the data. The logging program would also crash from time to time. This error could occur a few minutes or a few days after the logging program was switched on. Despite our best efforts we were unable to avoid this error which seems to have originated from the OS of the phone. On some of the phones when the program crashed an audible beep was made which reminded volunteers to turn on the program.

Due to the format in which the data was logged we were unable to ascertain the exact times for the occurence of these errors. However, we estimate from our data that on average the mobile probes were not logging for 24.5% of the time. From interviews with our probes, these outages seem to have been random and uniformly distributed over time. While we did miss potential contacts, our logs clearly mark the beginning and ending of any period when logging was performed. During these periods all potential contacts were recorded. In this paper we make inferences over these periods and hence the inferences about contact patterns are valid. Further, due to the random nature of the outages and the long interval over which we recorded data, the patterns that we look for in Sec. 5 are not significantly affected.

4 Metrics of Interest

The following definitions are crucial to understanding the metrics presented in this section. We define two devices to be in *contact* if they are in Bluetooth range of each other. We define a device to be in a *meeting*, if it is in *contact* with *at least* one other device for more than τ_c seconds.

Contact Time: This is the duration for which two bluetooth enabled devices are in contact. Contact time is a useful metric that spans all the applications that we have in mind. If contact times are longer on an average, it implies that more data can be exchanged/forwarded during each contact. For a DTN like application, this will affect the system throughput.

As has been established in earlier studies, the contact time distribution in Fig. 1(a) follows a power law relationship. 80% of the contacts are short in duration lasting less than 9 minutes. Contacts never last longer than 3 hours. The mean slope of the distribution is 0.84642, with a variance of 0.02 across the different mobile users. In other words, contact times are independent of user behavior.

Inter-pair-contact time: This is the time duration between two successive contacts of a specific pair of devices. If a pair of devices have been in contact only once, then the inter-pair-contact time is infinity. We look at inter-pair-contact times, conditioned on the event that this pair of devices has been in contact at least twice. This metric is of particular interest in DTN applications. For example in [3], they analyze the impact of the power law characteristic of inter-pair-contact time distributions on the performance of DTN.

The loglog plot for the inter-pair-contact time is shown in Fig. 1(b). Although the loglog plot looks linear over small time scales (within a day), it does not appear to follow a power law over large time scales. Note that in this study, there are no artifacts due to either granularity or time scale of the study, which were present in the Haggle and Toronto Bluetooth based studies. The best fit line to this curve has a slope of 0.414. We found that 80% of inter pair contacts occur within 2 hours. In this case, the slopes for individual users was quite different, which reflects individuality in user behavior. In [3], the authors found the inter-pair contact time to follow the power law with a coefficient of 0.6. We suspect the differences are due to the different environments encountered by the probes.

The two metrics above were investigated in previous DTN related studies. They focus on contacts between two specific people. However, not all applications can benefit from this information. An application like PeopleNet might be interested in just meeting any other users and swapping queries with them. It would benefit from knowing how and at what frequency people come together. The meeting metrics described below allow us to understand how people aggregate in groups.

Inter-contact time: This is the time between two successive contacts. For example if a device A is discovered at 1PM followed by a device B at 1 : 08 PM. Then, the inter-contact time is 8 minutes. This metric allows us to make inferences about the penetration of Bluetooth devices. It also determines the

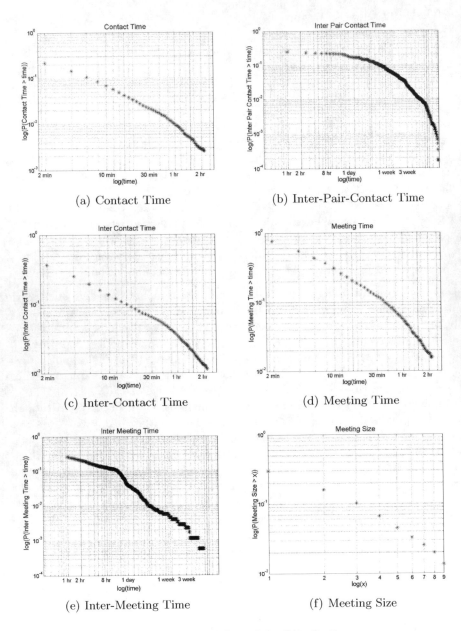

(a) Contact Time

(b) Inter-Pair-Contact Time

(c) Inter-Contact Time

(d) Meeting Time

(e) Inter-Meeting Time

(f) Meeting Size

Fig. 1. LogLog Plots of the Distribution

frequency with which one has opportunity to exchange data with devices in the proximity. This is key to the performance of PeopleNet and Serendipity like applications. In a PeopleNet like application, this metric will determine the time to generate a match between a pair of matching queries.

From Fig. 1(c), we see that the inter-contact time does follow a power law distribution. Inter-contact times never exceed a day in duration. The slope is 0.55. 80% of inter contact times occur within 40 minutes. However, if we constrain ourselves to consider only the work day, i.e. between 8AM and 8PM most of the inter-contact times were between 4 minutes to 13 minutes for various users.

Meeting Time: This is the duration for which a device is contiguously in a meeting state. We will illustrate this with an example. Suppose node N comes in contact with device A at 1PM and stays in contact for more than τ_c seconds, then N enters the meeting state at 1PM. Assume that N is in contact with A for one hour and B comes in contact at $1:05$ PM and stays in contact with node N until $2:05$ PM. At $2:05PM$, N is in contact with no other devices. Then we say that the meeting time is 65 minutes. Meeting times allow us to understand how people aggregate in groups. This is particularly relevant to PeopleNet like applications where this data could be used to estimate how long the application will have to swap queries with other users.

For distributions related to meetings, we computed them for different thresholds τ_c ranging from 30 seconds to 120 seconds. In the following plots, $\tau_c = 30$ seconds. The conclusions are very similar for the other threshold times also. We see from Fig. 1(d) that the distributions for the meeting times is well approximated by a power law. The slope for the meeting time is 0.76 and 80% of meetings last less than 30 minutes.

Inter-Meeting Time: This is the time interval between successive meetings. Again this is well approximated by a power law with a slope of 0.39. We note that there is a similarity between the distribution of inter-pair-contact times and inter meeting times.

Meeting Size: This is the total number of devices which are in contact over the duration of a meeting. This metric too follows a power law. The maximum meeting size we have discovered is 124. Interestingly, we noted that one can make predictions regarding the meeting time by knowing the meeting size. We found that if the meeting size was 1 the average meeting time was 13 minutes, whereas for meetings of size 2 and greater, the meetings lasted 17 minutes on average. Another interesting observation is that while 80% of the meetings are of size 1, 80% of all the contacts made are made in meetings of size 2 and above. This implies that aggregation centres, have the most potential to be exploited by the DTN applications.

Average Instantaneous Meeting Size: This average is computed for each meeting, by weighting each device in the meeting by the fraction of time for which that device was in contact with the probe. This helps us to understand aggregation dynamics. For any meeting, if the average is approximately the same as the meeting size, then the environment is fairly static. On the other hand, if the average is much less than the meeting size, this implies that the environment is dynamic with devices coming in and going out (e.g., in a mall).

We found that for most meetings, irrespective of the meeting size, the average instantaneous meeting size was close to 1. This implies that users are often discovered in dynamic environments such as malls and coffee shops and are constantly moving in and out of contact. Based on this we can make the conclusion, that whenever we make contact with a new device we must try and exchange information immediately, as the device might quickly leave the meeting.

5   Looking for Patterns

With the belief that predictability in contact patterns can aid algorithm design we looked for patterns in the following two ways.

5.1   Time Series

We analyzed individual user contact patterns and plotted time series at different time scales and looked for patterns in the number of distinct devices that the probes saw on varying time scales. When looking at a PeopleNet like application, it would be helpful to know when exactly a probe is likely to find herself in a crowded area which increases the opportunity to find matches for queries.

From Fig. 2(a), which captures the number of distinct devices seen by a certain probe on a particular day of the week, we see a clear diurnal pattern. There are no devices discovered between 9PM and 9AM for this probe. We observed similar pattern for most users. Typically, we found that no devices are discovered between midnight and 8 AM in the morning. We see see a fairly large variance in the number of devices discovered at any given hour.

Next we look at the number of contacts made by a user on each day of the week. This is shown in Fig. 2(b). We see that there is a large variance in the number of users seen on a given day making predictability hard. There also does not seem to be any clear difference between week days and weekends.

The time series data that we have looked at focuses on patterns in the number of contacts made. Consider, however, DTN applications where information needs to be passed from a particular source to a particular destination. An application such as this might not be so concerned with how many contacts will be made at different times, rather with whom those contacts might be and when they are likely to be made. To address this issue of estimating the chance of meeting specific people repeatedly we introduce the notion of commonality.

5.2   Commonality

For any given time scale, the commonality is computed by looking at all subsequent pairs of times. Assume we are considering times t_i and t_{i+1} at a particular time scale. Let A_i be the set of users seen at time t_i. Then the commonality for this pair of times is given by $C_i = \frac{|A_i \cap A_{i+1}|}{|A_i \cup A_{i+1}|}$. We then average over all i.

We first looked at the fraction of identical devices seen across subsequent hours on the same day (HSD). We found that the commonality ranged between 0.1 to 0.2. When we increased the time scale to subsequent days (SD) we found it had lesser commonality. For thoroughness, we also considered a daily time scale

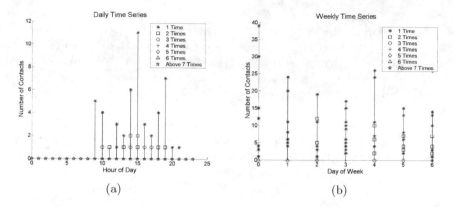

Fig. 2. Fig. 2(a) shows the number of contacts in each hour across a day on a particular day of the week. Fig. 2(b) shows the number of contacts made on every day of the week. 0 represents Sunday, 1 stands for Monday and so on. The different symbols are used to refer to the number of times the particular number of contacts found. For example, in Fig. 2(b), 0 devices were found 6 times on a monday and 30 devices were found once on a sunday.

to look at commonality between contacts at specific times on subsequent days (e.g. 10 AM everyday) (STSD), specific days of the week (DW) (e.g. all devices discovered on Thursdays) and at specific times on a weekly time scale (TDW) (e.g. devices discovered at 10 AM on Thursdays). We summarise the maximum commonality seen at all these time scales in table 2. The data indicates that the more time that has elapsed since the last time contact was made with a particular device, the lesser the chances of meeting that device again.

Table 2. Maximum commonality across different time scales. Refer paper for the meanings of acronyms. As time scale increases, there is reducing commonality between devices discovered.

Time Scale	Maximum Commonality
HSD	0.2
SD	0.14
STSD	0.07
DW	0.04
TDW	0.03

6   Discussion and Conclusions

In summary, we have performed an extensive data collection and analysis experiment with Bluetooth phones, getting a true sampling of user interactions. We believe that the data significantly builds on the existing data from other studies. In analyzing the data, apart from the usual metrics such as contact time and inter-contact times, we proposed several new metrics which help to understand specific behavior such as how users cluster in groups. We must note however, that the metrics are affected by the penetration of Bluetooth devices in the environment. We also looked for patterns in the interactions for the probes. In terms of the number of contacts seen there exists a certain amount of predictability. When we looked for correlations between the devices seen at different time scales, we found very little correlation.

In our future work an important area that needs attention is making the logging process more reliable. We have succeeded in modifying the program and making it far more stable. Another step we plan is to periodically text probes to keep their logging programs switched on.

References

1. P. Bahl A. Balachandran, G. Voelker and P. Rangan. Characterizing user behavior and network performance in a public wireless lan. In *In Proceedings of ACM SIGEMTRICS 2002*, Marina Del Ray, USA, June 2002.
2. T. Gross C. Tuduce. A mobility model based on wlan traces and its validation. In *Proceedings of Infocom 2005*, Miami, U.S.A, March 2005.
3. A. Chaintreau, P. Hui, J. Crowcroft, C. Diot, R. Gass, and J. Scott. Pocket switched networks: Real world mobility and its consequences for opportunistic forwarding. In *Technical Report Number 617*, February 2005.
4. N. Eagle and A. Pentland. Social serendipity: Mobilizing social software. *IEEE Pervasive Computing*, 4(2):28–34, 2005.
5. T. Henderson, D. Kotz, and I. Abyzov. The changing usage of a mature campus-wide network. In *Proceedings of MobiCom 2004*, pages 22–31, September 2004.
6. J. C. Cai J. Su, A. Chin and E. DeLara. User mobility for opportunistic ad-hoc networking. In *In Proceedings of IEEE WMCSA 2004*, English Lake District, UK, December 2004.
7. M. McNett and G. M. Voelker. Access and mobility of wireless PDA users. *Mobile Computing and Communications Review*, 9(2):40–55, 2005.
8. M. Motani, V. Srinivasan, and P. Nuggenhalli. PeopleNet: Engineering a wireless virtual social network. In *Proceedings of MobiCom 2005*, Cologne, Germany, August 2005.
9. M. Motani, V. Srinivasan, and O. Wei Tsang. Analysis and implications of student contact patterns derived from campus schedules. In *Proceedings of MobiCom 2006*, California, U.S.A., September 2006.
10. InfoComm Development Authority of Singapore. Telecom services statistics for may 2005. http://www.ida.gov.sg/idaweb/media/.
11. D. Tang and M. Baker. Analysis of a metropolitan-area wireless network. In *In Proceedings of ACM MOBICOM 1999*, Boston, USA, 1999.

Two Days in the Life of the DNS Anycast Root Servers

Ziqian Liu[2], Bradley Huffaker[1], Marina Fomenkov[1],
Nevil Brownlee[3], and kc claffy[1]

[1] CAIDA, University of California at San Diego
[2] CAIDA and Beijing Jiaotong University
[3] CAIDA and The University of Auckland
{ziqian,bhuffake,marina,nevil,kc}@caida.org

Abstract. The DNS root nameservers routinely use anycast in order
to improve their service to clients and increase their resilience against
various types of failures. We study DNS traffic collected over a two-day
period in January 2006 at anycast instances for the C, F and K root
nameservers. We analyze how anycast DNS service affects the worldwide
population of Internet users. To determine whether clients actually use
the instance closest to them, we examine client locations for each root
instance, and the geographic distances between a server and its clients.
We find that frequently the choice, which is entirely determined by BGP
routing, is not the geographically closest one. We also consider specific
AS paths and investigate some cases where local instances have a higher
than usual proportion of non-local clients. We conclude that overall,
anycast roots significantly localize DNS traffic, thereby improving DNS
service to clients worldwide.

Keywords: DNS, anycast, Root Servers, BGP.

1 Background

The Domain Name System (DNS) [1] is a fundamental component of today's
Internet: it provides mappings between domain names used by people and the
corresponding IP addresses required by network software. The data for this mapping is stored in a tree-structured distributed database where each nameserver
is authoritative for a part of the naming tree. The DNS *root nameservers* play
a vital role in the DNS as they provide authoritative referrals to nameservers
for generic top-level domains (gTLD, e.g. .com, .org) and country-code top-level
domains (ccTLD, e.g. .us, .cn).

When the DNS was originally designed, its global scope was not foreseen, and
as a consequence of design choices had only 13 root nameservers ("roots") that
would provide the bootstrap foundation for the entire DNS system. As the Internet grew beyond its birthplace in the US academic community to span the world
it increasingly put pressure on this limitation, at the same time also increasing
the deployment cost of any transition to a new system. Thus, anycast [2] was

S. Uhlig, K. Papagiannaki, and O. Bonaventure (Eds.): PAM 2007, LNCS 4427, pp. 125–134, 2007.

presented as a solution since it would allow the system to grow beyond the static 13 instances, while avoiding a change to the existing protocol. For a DNS root nameserver, anycast provides a service whereby clients send requests to a single address and the network delivers that request to at least one, preferably the closest, server in the root nameserver's anycast group [3].

We define an *anycast group* as a set of instances that are run by the same organisation and use the same IP address, namely the *service address*, but are physically different nodes. Each instance announces (via the routing system) reachability for the same prefix/length – the so-called *service supernet* – that covers the service address and has the same origin Autonomous System (AS). The service supernet is announced from different instances by Border Gateway Protocol (BGP) such that there may be multiple competing AS paths. Instances may employ either *global* or *local* routing policy. Local instances attempt to limit their *catchment area* to their immediate peers only by announcing the service supernet with `no-export` attribute. Global instances make no such restriction, allowing BGP alone to determine their global scope, but use prepending in their AS path to decrease the likelihood of their selection over a local instance [4].

As of today, anycasting has been deployed for 6 of the 13 DNS root nameservers, namely, for the C, F, I, J, K and M roots [5]. The primary goal of using anycast was to increase the geographic diversity of the roots and isolate each region from failures in other regions; as a beneficial side effect, local populations often experience lower latency after an anycast instance is installed. As well, anycast makes it easier to increase DNS system capacity, helping protect nameservers against simple DOS attacks. The expected performance gains depend on BGP making the best tradeoff between latency, path length and stability, and Internet Service Provider (ISP) cost models. BGP optimizes first ISP costs and then Autonomous System (AS) path length, attaining any gains in latency and stability as secondary effects from this optimization.

In this study we examine traffic at the anycast instances of the C, F, and K root nameservers and their client population. We substitute the geographic proximity as a proxy for latency, since latency between metropolitan areas is dominated by propagation delay [6].

2   Data

Measurements at the DNS root nameservers were conducted by the Internet Systems Consortium (ISC) and the DNS Operations and Analysis Research Center (OARC) [7] in the course of their collaboration with CAIDA. DNS-OARC provides a platform for network operators and researchers to share information and cooperate, with focus on the global DNS.

The full OARC DNS anycast dataset contains full-record `tcpdump` traces collected at the C, E, F, and K-root instances in September 2005 and January 2006. The traces mostly captured inbound traffic to each root instance, while a few instances also collected outbound traffic. For this study we selected the most complete dataset available, the "OARC Root DNS Trace Collection January

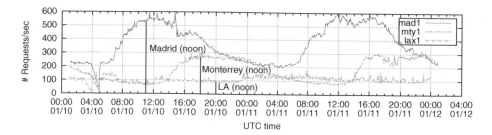

Fig. 1. Diurnal patterns of the DNS traffic to the F-root local instances `mad1` (Madrid, Spain), `mty1` (Monterrey, Mexico), and `lax1` (Los Angeles, US). For each instance, the local time noon is explicitly specified with a solid vertical line. The artifact on Jan. 10th between 4:00 and 5:00 appears because no data available for this period.

2006" [8]. It includes traces collected concurrently at all 4 C-root instances, 33 of the 37 F-root instances and 16 of the 17 K-root instances during the period from Tue Jan 10 to Wed Jan 11 2006, UTC. A common maximum interval for all measured instances is 47.2 hours or nearly two whole days.

Each of the three root nameservers we measured implements a different deployment strategy [9]. All nodes of C-root are routed globally, making its topology flat. The F-root topology is hierarchical: two global nodes are geographically close, with many more widely distributed local nodes. Finally, K-root represents a case of hybrid topology with five global and 12 local nodes, all geographically distributed. The instance locations for all roots are listed in [5].

Our target data are IPv4 UDP DNS requests to each root server's anycast IP address. Some of the F and K-root instances have applicable IPv6 service addresses, and we observed a few requests destined to these addresses. Further analysis of the IPv6 DNS traffic is needed, but in this paper we focus on IPv4 traffic. We also note that for the F and K-root instances that collected TCP traffic associated with port 53, its volume was negligible, namely, ~1.3% of total bytes and ~3.2% of total packets.

3 Traffic Differences Between Root Server Instances

3.1 Diurnal Pattern

Assuming that DNS traffic is primarily generated by humans, rather than by machines, we expect to see a clear diurnal pattern for those instances that primarily attract a client base from a small geographic area. Fig. 1 shows the time distribution of DNS requests to three F-root local instances: `mad1`, `mty1` and `lax1`. Both `mad1` and `mty1` have a clear diurnal pattern matching the local time, i.e. rising in the morning and falling towards midnight. However, `lax1` has a distinct traffic pattern, where the crest of the request curve is shifted from its local midday by ~8 hours. This difference suggests that a large proportion of `lax1`'s requests are coming from clients who do not follow the local time

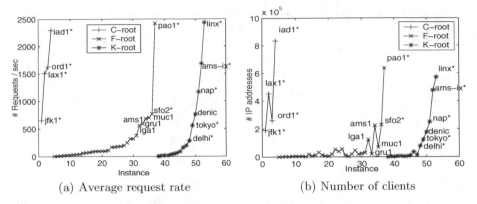

(a) Average request rate   (b) Number of clients

Fig. 2. Average instance requests per second and the total number of clients.
The x-axis instance order is the same in both (a) and (b). The instances are plotted
in groups for C, F and K roots; within each group they are arranged in an increasing
request rate order. Symbol * designates global instances.

of the instance, most likely, because they are located elsewhere. Indeed, as we
show in Section 4.1, although lax1 is located in the US, ∼90% of its clients are
in Asia and they generated over 70% of the total requests that this instance
received.

We also studied the request time distribution of one of the global instances
(not shown) and found that its curve was flatter than those of local instances.
However, slight diurnal variations were still noticeable and correlated with the
local time of the continent from which that global instance has the largest proportion of its clients.

3.2   Traffic Load

We characterised the traffic load of root server instances with two metrics: number of requests per second averaged over our measurement interval and total
number of clients served during this interval (Fig. 2). Global instances generally
have higher request rates and serve larger populations than local instances, but
there is large variability in their loads. Some local instances also have fairly high
traffic loads and large client populations comparable to those of the global instances. Such high loads may occur because (1) the local instance's catchment
area has a high density of Internet users that generate many requests, or (2)
its catchment area is topologically larger than normal. For example, the F-root
local instance ams1 is peering with AMS-IX, an Internet exchange point in Amsterdam, NL, which is one of Europe's major exchange points. Therefore, ams1
peers with a large number of ASes via AMS-IX and attracts a higher request
rate and larger number of clients than is typical for a local instance. At the same
time, some local instances have extremely low load levels (less than 10 pkt/s on
average over two days period), serve only a handful of clients, and are clearly
underutilised.

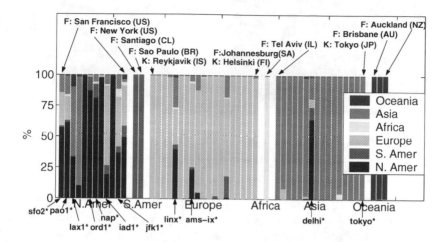

Fig. 3. Client continental distribution of instances. Each bar represents one instance, and the bars are arranged from left to right according to the instance longitude, in the west to east order. Groups delimited by white gaps represent instances located in the same continent. The anycast group (root) and the city names of the instances that are located at continent boundaries are given above the bars. Within each bar, the colored segments show the distribution of clients by continent. Global instances are marked below the bars, where the first row is for F-root, the second row is for K-root, and the third row is for C-root. To conserve space the legend overlaps some bars, but the bar color does not change within the overlapped area.

The non-monotonically increasing curves in Fig. 2(b) indicate that the number of requests to a server can be disproportional to the number of clients it serves. Classification of users as "heavy" and "light" and a detailed analysis of their behavior patterns is a subject of future research.

4   Anycast Coverage

4.1   Client Geographic Distribution

To discover the geographic distribution of each instance's clients, we map the client IP addresses to their geographic locations (country and continent) and coordinates (latitude and longitude) using Digital Envoy's NetAcuity database [10]. The database claims accuracy rates over 99% at the country level and 94% at the city level worldwide.

C and F-root instances are named using their corresponding airport codes, e.g. f-lax1 denotes the F-root instance at Los Angeles, while K-root instances are named either after the exchange points that support them, or their city name. Therefore, for the root server instance locations we use the coordinates of the closest airport. We then compute the geographic distance between instances and their clients as the great circle distance.

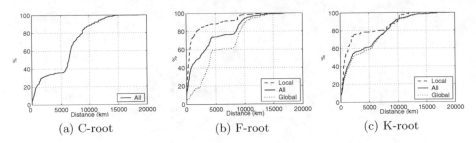

(a) C-root   (b) F-root   (c) K-root

Fig. 4. CDF of the distance from root nameservers to their clients

Continental Distribution. Distribution of clients by continents for each measured instance is shown in Fig. 3. Comparing clients of local and global instances we notice that clients of most global instances are indeed distributed worldwide. For example, the K-root global instance linx located at London had only 28.6% of its total clients from Europe. Others were from: North America 40%, South America 3.7%, Africa 1.6%, Asia 24.3%, and Oceania 1.8%. Most of the local instances were serving clients from the continent they are located in. For example, nearly all the clients of the F-root local instance at Santiago, Chile are from South America. Such a constrained geographical distribution is consistent with the goal of DNS anycast deployment: to provide DNS root service closer to clients.

There are exceptions among both global and local instances. Over 99.7% of K-root's tokyo instance were from Asia. Furthermore, 75% of its clients were less than 1000 km away, i.e. mostly in Japan. Hence, this instance behaves more like a local instance rather than a global one. The previously mentioned F-root local instance lax1 at Los Angeles, US (the 4th bar from the left, not to be confused with the C-root global instance lax1* which has the same code name) has 88% of its clients from Asia, and only 10% from North America, which explains its irregular diurnal pattern in Fig. 1. Such abnormal client distributions result from the instances' BGP routing configurations, which we discuss in Section 4.2.

Distance Distribution. We also study the distance from root server instances to their clients. Fig. 4 plots, for each root server, a CDF for its local instances, its global instances, and all its instances combined Only one curve is given for the C-root instances since they are all global.

Fig. 4 shows that the majority of the local instances were serving clients who are geographically close to them – 80% of the F-root local instances' clients and 70% of the K-root local instances' clients were within 1800 km. Distances between the global instances and their clients are generally longer, e.g. for C-root, over 60% of the clients were beyond 5000 km, and F- and K-root both had 40% of their clients beyond 5000 km. The F and K roots had lower proportions of clients who were far away from their servers because these anycast groups include multiple local instances all over the world while the C-root group currently has only 4 instances and they are all global.

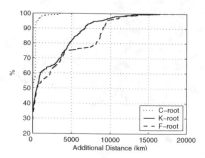

Fig. 5. CDF of additional distance travelled by requests to instances

Flat segments in the CDF curves (around 5000 km for the C and K roots, and the especially prominent one from 5000 to 8000 km for the F-root) approximately correspond to the distances across the Atlantic Ocean (from North America to Europe) and the Pacific Ocean (from North America to Asia), respectively. Obviously, fewer clients are found in the ocean areas.

Additional Distance. We wanted to investigate whether the BGP always chooses the instance with the lowest delay. For this analysis, we use the geographic proximity as a proxy for latency. A comprehensive study [6] shows that geographic distance usually correlates well with minimum network delay. Later studies [11,12] also used geographic distance to compute network delay.

For a given client, we define the *serving instance* as the instance the client actually uses, and the *optimal instance* as the geographically closest instance from the same anycast group. We ignore the tiny number of clients that sent requests to more than one instance. (see Section 4.3 below). We then define the client's *additional distance* as the distance to its serving instance minus the distance to its optimal instance. An additional distance of zero indicates that the client queried an optimal instance while a positive value suggests a possible improvement.

Analyzing the CDF of the additional distance (Fig. 5), we saw that 52% of C-root's clients were served by their optimal C-root instance, and another 40% had short additional distances. This optimised selection is due to the flat topology of the C-root anycast group, i.e., all instances are global. In contrast, only 35% of F-root's clients and only 29% of the K-root's clients were served by their optimal instances. Given that the speed of light in fiber is about 2×10^8 m/s, an additional 5000 km of geographical distance adds a 25 ms delay. Our results imply that a significant number of clients would benefit if routing configurations of their local DNS root instances were optimized to route these clients to their optimal instance, thereby reducing their DNS service delay.

4.2 Topological Coverage

We studied the topological coverage of the Internet by anycast clouds of the C, F, and K root nameservers. Using the RouteViews BGP tables [13] collected on 10 Jan 2006, we mapped each client IP address in our data to its corresponding

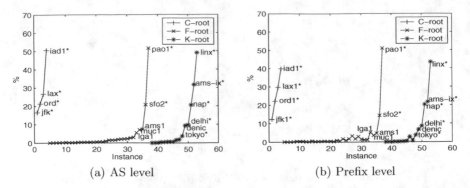

(a) AS level   (b) Prefix level

Fig. 6. Topological scope of the instances. The x-axis instance order is the same in (a) and (b). The instances are plotted in groups for C, F and K roots; within each group they are arranged in an increasing AS coverage percentage order. The percentage shown for of each instance is the number of ASes/prefixes seen by the given instance divided by the total number of ASes/prefixes seen by all three roots combined.

prefix by longest matching, and so determined its origin AS. Out of 21883 ASes seen in RouteViews tables on that day, we observed IP addresses belonging to 19237 ASes ($\sim$88%) among our clients.

Fig. 6 shows both the AS-level and prefix-level coverage for each instance relative to the total number of ASes (prefixes) seen by all instances of the three root nameservers. As expected, most of the global instances have much higher topology coverage than the local instances.

Two exceptions are: (1) the K-root local instance `denic` in Frankfurt, Germany had a wider topological scope than any other local instances; (2) the K-root global instance `tokyo` saw a rather small fraction of ASes and prefixes. Such exceptions can be explained by RouteViews BGP data.

Knowing the IP address of the K-root anycast service supernet and using the AS peering information published at the root server website [14], we extracted the AS paths to each of its instances. One of the three observed AS paths to `denic` is `12956 8763 25152`. According to the RIPE-NCC *whois* database, AS12956, belongs to Telefonica, which has a global network infrastructure. The presence of this path explains why `denic` has a high topological coverage, and, correspondingly, a high traffic load and a large number of clients (cf. Fig. 2).

Considering the K-root instance `tokyo`, we note that the global instances used AS-path prepending to intentionally lengthen their paths. This instance announced a triple AS-prepended path, i.e. `4713 25152 25152 25152 25152` which was the longest among all of the five K-root global instances. Such a long AS path caused `tokyo` to be seldom chosen by BGP for global clients who sent queries to the K-root server. Therefore, the clients of `tokyo` were mostly local (cf. Fig. 3).

Finally, we saw in Section 4.1 that the F-root local instance `lax1` had most of its clients coming from Asia. One of the three AS paths we observe to this instance was: `7660 2516 27318 3557`, where AS7660 and AS2516 are in Japan thus explaining the source of the Asian clients.

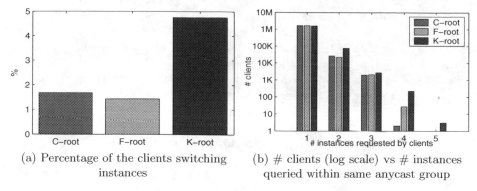

(a) Percentage of the clients switching instances

(b) # clients (log scale) vs # instances queried within same anycast group

Fig. 7. Instance switching within the same anycast group

4.3 Instance Affinity

Anycast improves stability by shortening AS paths, thus decreasing the number of possible failure points. However, this enhancement comes at the cost of increased chance of inconsistency among instances [2] and of clients' transparent shifting to different instances. As long as DNS traffic is dominated by UDP packets, this route flapping is unimportant, but it may pose a serious problem if stateful transactions such as TCP or multiple fragmented UDP packets become more prominent [15]. Fortunately, recent studies [9,16,17] suggest that the impact of routing switches on the query performance is rather minimal.

We observed that a small fraction of clients did switch instances during the two days: 1.7% of the C-root clients, 1.4% of the F-root clients, and 4.7% of the K-root clients (Fig. 7(a)). These percentages correlate with the number of global instances each root server has (4 for C-root, 2 for F-root, and 5 for K-root), since the clients of a global instance are more easily affected by routing fluctuations. Actually, the two F-root global instances together saw approximately 99.8% of the total clients who switched F-root instances, and the five K-root global instances together saw 86% of the total clients who switched K-root instances.

Fig. 7(b) shows how many clients queried how many instances. Focusing on the clients who used the most instances, we found that the two C-root clients who requested four instances were from Brazil and Bolivia and the three K-root clients who requested five instances were all from Uruguay. Note that neither the C-root nor the K-root had an instance in South America. For F-root, the 27 clients who requested four instances were all from the UK where the F-root has a local instance lcy1, but the catchment area of this instance was limited. Actually, those 27 clients never requested from lcy1, but switched between ams1, lga1, pao1, and sfo2. A detailed analysis of unstable clients could help network designers decide where to place new instances.

5   Conclusion

From the diurnal patterns of request rates and from the observed geographic clustering of clients around instances we conclude that the current method for limiting the catchment areas of local instances appears to be generally successful. A few exceptions, such as the F-root local instance `lax1` or the K-root local instance `denic`, drew their clients from further away regions due to peculiar routing configurations.

Instance selection by BGP is highly stable. Over a two-day period less than 2% of both C-root and F-root clients and <5% of K-root clients experienced an instance change. Since UDP connections are stateless, the vast majority of clients would not be harmed by such changes, apart from the unavoidable delay created by BGP convergence. Although the instance flapping could be problematic to TCP's stateful connections [15], in our data sample TCP packets constituted only 3.2% of all DNS root packets.

Overall, the transition to anycasting by the DNS root nameservers not only extended the origial design limit of 13 DNS roots, but it also provides increased capacity and resilience, thereby improving DNS service worldwide.

Acknowledgements. We thank ISC, RIPE, and Cogent for collecting the datasets used in this study. P. Vixie, K. Mitch, and B. Watson of ISC helped with data storage and answered questions on F-root's anycast deployment. A. Robachevsky and C. Coltekin from RIPE provided feedback on K-root's anycast deployment. This work was supported by NSF Grant OCI-0427144.

References

1. Mockapetris, P.: Domain Names - Concepts and Facilities, Internet Standard 0013 (RFC 1034, 1035), Nov. 1987.
2. Hardie, T.: Distributing Authoritative Nameservers via Shared Unicast Addresses. RFC 3258, Apr. 2002.
3. Partridge, C., Mendez, T., Milliken, W.: Host Anycasting Service. RFC 1546, 1993.
4. Abley, J.:Hierachical Anycast for Global Service Distribution.
   http://www.isc.org/pubs/tn/isc-tn-2003-1.html
5. DNS root nameservers web sites. http://www.root-servers.org/
6. Padmanabhan, V.N., Subramanian, L.: An Investigation of Geographic mapping techniques for Internet Hosts. ACM SIGCOMM, Aug. 2001.
7. OARC. https://oarc.isc.org/docs/dns-oarc-overview.html
8. OARC Root DNS Trace Collection January 2006 http://imdc.datcat.org/collection/1-00BC-Z=OARC+Root+DNS+January+2006
9. Colitti, L., Romijn, E., Uijterwaal, H., Robachevsky, A.: Evaluating The Effect of Anycast on DNS root nameservers. Unpublished paper, Jul. 2006.
10. NetAcuity. http://www.digital-element.net
11. Spring, N., Mahajan, R., Anderson, T.: Quantifying the Causes of Path Inflation. ACM SIGCOMM, Aug. 2003.
12. Sarat, S., Pappas, V., Terzis, A.: On the Use of Anycast in DNS. ACM SIGMETRICS, Jun. 2005
13. Route Views Project. http://www.routeviews.org
14. K-root Homepage. http://k.root-servers.org/
15. Barber, P., Larson, M., Kosters, M., Toscano, P.: Life and Times of J-root. NANOG 32, Oct. 2004
16. Boothe, P., Bush, R.: DNS Anycast Stability.19th APNIC, Feb. 2005.
17. Karrenberg, D.:Anycast and BGP Stability.34th NANOG, May 2005.

The Internet Is Not a Big Truck: Toward Quantifying Network Neutrality[*,**]

Robert Beverly[1], Steven Bauer[1], and Arthur Berger[2]

[1] MIT CSAIL, Cambridge MA 02139, USA
[2] Akamai/MIT CSAIL, Cambridge MA 02139, USA
{rbeverly,bauer,awberger}@csail.mit.edu

Abstract. We present a novel measurement-based effort to quantify the prevalence of Internet "port blocking." Port blocking is a form of policy control that relies on the coupling between applications and their assigned transport port. Networks block traffic on specific ports, and the coincident applications, for technical, economic or regulatory reasons. Quantifying port blocking is technically interesting and highly relevant to current *network neutrality* debates. Our scheme induces a large number of widely distributed hosts into sending packets to an IP address and port of our choice. By intelligently selecting these "referrals," our infrastructure enables us to construct a per-BGP prefix map of the extent of discriminatory blocking, with emphasis on contentious ports, i.e. VPNs, email, file sharing, etc. Our results represent some of the first measurements of network neutrality and aversion.

1 Introduction

As the Internet has matured, its success has spurred not only technical innovation, but also social, economic and regulatory responses [1]. One initially unanticipated response is a form of policy control employed by network operators known as "port blocking." Port blocking relies on the close coupling between particular applications and their assigned TCP or UDP port. Since many applications use well-known port numbers [2,3], port blocking is one technique to stop traffic belonging to a particular application or class of application.

This research seeks to quantify the extent of port blocking on the Internet. We present a hybrid active/passive measurement-based approach that is capable of rapidly testing large parts of the Internet topology. Our scheme induces peer-to-peer (P2P) clients in the Gnutella network to probe for port blocking as part of their natural overlay formation process. Our technique does not degrade or disrupt the performance of the P2P network.

Our objective is to provide unbiased information about port blocking on the Internet. We therefore do not attempt to argue which network operational practices are "legitimate" or "justifiable." Such judgments are not purely technical, but rather must be made in the context of a well-informed larger discussion.

---

[*] Flippant title adopted from Senator Ted Stevens' remarks to the United States Senate Commerce Committee vis-à-vis network neutrality.

[**] This work supported in part by Cisco Systems and NSF Award CCF-0122419.

S. Uhlig, K. Papagiannaki, and O. Bonaventure (Eds.): PAM 2007, LNCS 4427, pp. 135–144, 2007.
© Springer-Verlag Berlin Heidelberg 2007

The prevalence and type of port blocking is of technical interest to application developers and academics but, perhaps more importantly, has prominently arisen in regulatory and policy debates – in particular debates over **network neutrality** [4,5]. For example, the United States FCC recently ordered a provider to cease port blocking a competing telephony service [6].

> *"Allowing broadband carriers to control what people see and do online would fundamentally undermine the principles that have made the Internet such a success."* – Vint Cerf [7]

Unfortunately, many underlying arguments that guide neutrality discussions are based on assumptions rather than careful measurement. While many definitions of neutrality exist, port blocking is an important and well-defined dimension of the debate. Port blocking is a simple and cheap mechanism for operators to control the type of traffic on their network. Indeed blocking can be employed for altruistic reasons, for instance to staunch the spread of Internet worms [8], or as a security measure to protect potentially vulnerable applications [9]. However, providers also leverage blocking for anti-competitive or economic purposes, for example blocking high-bandwidth file sharing applications or forcing subscribers to use their provider's email gateway [10,11,12,13].

Our primary contribution is the design and implementation of a novel methodology for measuring Internet port blocking. Based on initial measurements collected to date, the methodology seems to hold significant promise for systematic large-scale measurement of the port blocking dimension of network neutrality.

2   Measuring Port Blocking

In designing a methodology for measuring the extent and nature of Internet port blocking, we first examine what such measurements should include:

- *Generality:* Test any arbitrary port number in the 16-bit range allocated to the TCP and UDP protocols, i.e. $(0, 2^{16} - 1]$. Several ports bear special notice such as HTTP (port 80), SMTP (port 25), P2P file sharing, virtual private networks and games. These applications are some of the most contentious.
- *Range:* Test a wide range of networks across the entire Internet.
- *Quantity:* Test a large number of hosts across the entire Internet.
- *Minimal Participation:* Assume no active, coordinated or cooperative participation from remote hosts.

Active client participation, such as that used in the Spoofer Project [14] or the IPPM metrics [15], would enable us to comprehensively test a wide range of ports and even quality of service properties. However, per our last requirement above, we cannot assume active participation since it is at odds with testing a large quantity and range of networks. The seemingly difficult problem is to induce hosts, randomly distributed on the Internet, to send packets to a destination and port of our choice. Our approach uses clients participating in the Gnutella P2P file sharing overlay in a novel manner to accomplish the aforementioned goals.

2.1   Functional Overview

Unstructured overlays such as Gnutella allow nodes to interconnect with minimal constraints. To scale, they rely on a two-level hierarchy of leaves and "Super-Peers" [16]. The Gnutella overlay is formed organically with SuperPeers actively managing the number of connections they maintain both to other SuperPeers and to leaves [17]. A peer can turn away connection requests via a busy message. The busy response also includes other peers to try so that new nodes can bootstrap. Nodes successively attempt connections to peers until they find a stable set of links. Our system crucially relies on the fact that this busy "referral" includes both the IP address and port number of other peers to contact.

Figure 1 depicts the high-level architecture of our system (the complete system is described in §2.4). We manage two separate machines, a Rogue SuperPeer (RSP) and a measurement host. The RSP joins the Gnutella SuperPeer mesh and routes queries and responses according to the normal Gnutella protocol. Once connected, the presence of our RSP is advertised by other SuperPeers. When new leaf node clients attempt to connect to our RSP (step 1), it sends a busy message and deterministically advises the client to try connecting to our measurement host (step 2). In this fashion, we have effectively tricked the client into sending a packet to the IP and port number of our choosing (step 3).

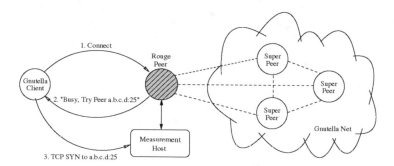

Fig. 1. Methodology: The Rogue SuperPeer joins the Gnutella network. (1) Client attempts to connect; (2) RSP rejects, referring the client to a measurement host under our control; (3) By correlating connections, we map Internet port blocking.

Any distributed system which allows arbitrary redirection messages is suitable for our task, for instance Bittorrent, HTTP links, etc. However, we use Gnutella as it easily facilitates global advertisement of the presence of our RSP. Additionally, the size and scope of the Gnutella network, approximately 2 million users [18], allows our method to elicit a high number of connections and thus redirect them for measurement purposes. Note that the initial SYN sent by clients does not contain data and hence is not affected by middle boxes that might drop packets based upon deep packet inspection.

2.2   A Map of Internet Port Blocking

Consider a client c residing on network W, i.e. c's IP address belongs to W. If c follows the busy referral from our RSP and connects on port p, we conclude that W does not block p (and thus is neutral to applications that use port p). However, the client c may not follow the referral or attempt the connection. Our measurement host must disambiguate whether the absence of a connection from c implies that W blocks p, or c never attempted to connect.

By intelligently selecting p in the busy redirect message of step 2 in Figure 1 on the basis of the client's network W, we overcome this ambiguity. We use a BGP routing table [19] to associate the client's IP address with BGP prefix b in the set of advertised prefixes B (i.e. a map of $client's\ IP \mapsto b \in B$). Once our measurement host receives a successful attempt from a (p, b) pair, the RSP does not attempt to test p for any future clients connecting from b. Next, we detail the system's port selection process in the face of uncertainty.

2.3   Probabilistic Inference

If a particular client does not heed the busy referral message, probabilistically the system will encounter another client that does. In the limit, our measurements can construct an accurate picture of the extent of discriminatory network port blocking. To formalize the conditions under which there is a high probability that a given network is blocking traffic, we first give the definitions in Table 1.

Table 1. Formal Definitions for Blocking Inference

$f(IP) = b \in B$	A function $f()$ on an IP address IP that gives an identifier b in the set of all BGP prefixes B
$P = \{p \mid p \in \mathbb{N}, 0 < p < 2^{16}\}$	The set P of all possible TCP or UDP ports
$n(p, b) \in \{0, 1\}$	A binary indicator variable. Is "1" (respectively "0") if IP traffic with destination port p is allowed (respectively blocked) on the path from originating BGP prefix b to the measurement host.
$H(p, b, i) \in \{0, 1\}$	A binary indicator variable. Is "1" (respectively "0") if the measurement host observed (respectively did not observe) a packet destined to port p from any of i clients with IP addresses in prefix b.

Given that the measurement host observes a packet, $H(p, b, i) = 1$, we trivially conclude that traffic to port p is allowed: $P(n(p, b) = 1 \mid H(p, b, i) = 1) = 1$. Not as trivial is the probability that traffic to port p is blocked, given that no packet was observed, $P(n(p, b) = 0 \mid H(p, b, i) = 0)$. By Bayes' Theorem:

$$P(n(p,b)=0 \mid H(p,b,i)=0) = \frac{P(H(p,b,i)=0 \mid n(p,b)=0)P(n(p,b)=0)}{\sum_{j=\{0,1\}} \left(P(H(p,b,j)=0 \mid n(p,b)=j) P(n(p,b)=j) \right)}$$

$$(1)$$

Since no packet will be observed if indeed traffic to the port is blocked, we have that $P(H(p, b, i) = 0|n(p, b) = 0) = 1$. Empirically (see §3.1), we find that the probability that a Gnutella client does not use the p reference the RSP passes along is approximately 0.8, which we conservatively estimate as 0.9. Assuming independence across the i clients, the probability no packet is observed if indeed traffic to the port is allowed is 0.9^i, i.e. $P(H(p, b, i) = 0|n(p, b) = 1) = 0.9^i$. Prior to our observations, we assume no information as to whether the port is blocked, and equal prior probabilities: $P(n(p, b) = 1) = P(n(p, b) = 0) = \frac{1}{2}$. [1] Substituting into (1), we obtain the probability that traffic to port p is blocked given that no packet was observed:

$$P(n(p, b) = 0|H(p, b, i) = 0) = \frac{1}{1 + 0.9^i} \tag{2}$$

$$\approx 1 - 0.9^i \quad \text{for } 0.9^i \quad \text{small} \tag{3}$$

We wish to set i such that if our measurement host does not receive a packet pair (p, b) after i redirect messages to hosts residing in prefix b, then the probability is suitably large that port p is indeed blocked. Choosing i such that

$$P(n(p, b) = 0|H(p, b, i) = 0) = 0.995 \Rightarrow i = \log_{0.9}(0.005) \simeq 50 \tag{4}$$

Thus, we must send ≈ 50 referrals to b for p to conclude, with probability 0.995, that port p is blocked on the path from b to the measurement host □.

2.4   Full Methodology Design

Based on the prior discussion, we present the full system methodology in Figure 2, an augmented version of Figure 1. All state is maintained in a database. The RSP and measurement hosts asynchronously read and write to the database to update the current state. The database also facilitates later off-line analysis.

Both the RSP and measurement host interface with a BGP database, built from a routeviews [19] table, that provides a mapping between an IP address and the longest matching prefix to which that address belongs. Each unique prefix is assigned a unique identifier in the database.

The "NextPort updater" is a process which runs every five minutes. The updater implements the logic in (§2.3) to intelligently update the database's notion of which port the RSP gives out in the next referral for a particular prefix in order to glean the most information. The updater orders the choice of p according to those most likely to be blocked, e.g. VPNs, file sharing, etc. Appendix A gives a complete description of the ports we explicitly test.

Lastly, the measurement host implements a front-end multiplexer which transparently redirects traffic from any incoming port to the port on which the SuperPeer is listening. In this fashion, clients connect to an actual SuperPeer irrespective of the port in the RSP's referral messages.

---

[1] One could choose to assume prior information. Suppose $P(n(p, b) = 1)$ is set equal to δ, and $P(n(p, b) = 0) = 1 - \delta$. Then equation (2) becomes $P(n(p, b) = 0|H(p, b, i) = 0) = \frac{1}{1 + 0.9^i * \delta/(1-\delta)} \approx 1 - 0.9^i * \delta/(1 - \delta)$. And (4) becomes $i = \log_{0.9}(0.005 * (1 - \delta)/\delta)$.

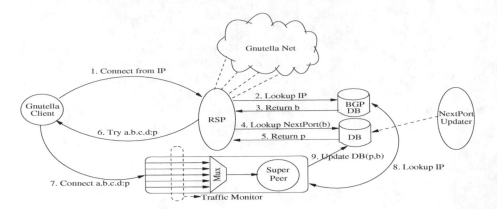

Fig. 2. Full Port Blocking Measurement Methodology

3   Results

We collected data using our infrastructure over two months in order to validate the methodology, refine testing and gather results. While our experiment is on-going, these initial results are very promising. The anonymized data from this study is publicly available from: http://ana.csail.mit.edu/rsp.

3.1   Efficacy of Methodology

The efficacy of our methodology depends firstly on issuing referrals to many Gnutella clients distributed across many networks. As seen in Table 2, over two months our RSP sent approximately 150k referrals to 72k unique Gnutella clients. These clients represent some 31k different global BGP prefixes, a non-trivial fraction of the Internet.

Table 2. Collection Statistics, Period: 02-Oct-2006 to 02-Dec-2006

	Count	Rate
Unique BGP Prefixes	31,219	0.7/Minute
SYN Packets Received	973,865	21.0/Minute
Unique IP Sources	328,437	7.4/Minute
Unique Gnutella Peers	72,544	1.6/Minute
Referrals Sent	147,581	3.3/Minute

Second, Gnutella clients which receive the specially crafted referrals from the RSP must follow the referral, i.e. attempt a connection on the basis of the referral, a non-negligible fraction of the time. Since a Gnutella client attempts only to find a stable set of connections into the network, it is unsurprising that not all potential SuperPeers are explored. We observe variability in the fraction of referrals that clients follow. Figure 3 depicts the fraction of followed referrals versus the cumulative fraction of clients.

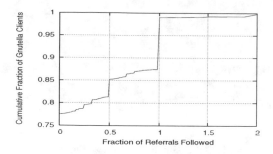

Fig. 3. RSP Referral Efficacy: proportion of referrals followed vs. cumulative fraction of clients

Fully 78% of the clients to which which our RSP sends referrals never result in a connection attempt. However, approximately 5% of the clients take half of all referrals and another 10% follow all referrals. Manual inspection of the clients which follow all referrals suggests that these clients are actually Gnutella network spiders. Because the spiders attempt to search and index the network, they follow all possible links in the overlay.

Thus, our referral methodology operates exactly as anticipated and allows us to build a map of port blocking given a sufficiently large collection window. The measured 78% non-connection attempt rate corresponds directly to the conditional probability of a Gnutella client not following an RSP reference from equation (2): $p(H(p, b, 1) = 0|n(p, b) = 1) = 0.78$.

Note that we record incoming connection attempts from clients the RSP has not interacted with and ports for which the RSP has not handed out referrals. On inspection, these connections appear to be from malicious hosts and malware performing random scanning. As this connection information is in some sense additional data for free, we include it in our analysis.

3.2 Observed Port Blocking

Given the approximately 1M incoming SYN packets observed by our measurement SuperPeer and induced by our RSP, we can begin to make per-BGP prefix inferences of port blocking. In this initial analysis we restrict our definition of blocking to blocking at any point along the path from the client to our servers; in future work we plan to use additional techniques to understand individual autonomous system behavior. Of the 31,000 prefixes, we find 256 prefixes which exhibit blocking for one or more ports as determined by Equation (4). Let α_p be the ratio of number of inferred prefixes blocking p to the total number of prefixes for which our measurement host has classified. Let $\#\{A\}$ denote the number of elements in set A. Then formally:

$$\alpha_p = \frac{\#\{b \text{ such that } n(p, b) = 0\}}{\#\{b \text{ such that } n(p, b) = 0\} + \#\{b \text{ such that } n(p, b) = 1\}} \tag{5}$$

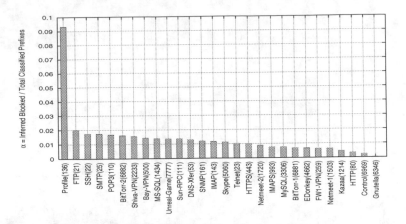

Fig. 4. Per-port α: blocked versus total inferences over observed BGP prefixes

Figure 4 shows the relative incidence of port blocking by giving p versus α_p for $\alpha_p > 0 \; \forall p$. We use $p = 6969$ as a control group as this port is unassociated with any applications or vulnerabilities and is typically unblocked.

We highlight only the most interesting results due to space constraints. The most frequently blocked port is 136, the collateral damage port which we discuss in (§4). The three lowest α_p in descending order are HTTP, Control and Gnutella which matches our intuitive notion of commonly open ports and serves as a methodology litmus test. Email ports (25, 110, 161, 143) are more than twice as likely to be blocked as our control port. Port 1434 was widely blocked due to a worm [8] and shows prominently three years after the initial outbreak. FTP, SSH, Bittorrent and VPNs round out the remaining top blocked ports.

Manual inspection of the BGP prefixes to which blocking is attributed reveals several ISPs and universities blocking outgoing P2P ports (1214, 4662, 6346, 6881). We find Canadian and US ISPs as well as a Polish VoIP prefix blocking Skype. Email, especially outbound port 25 (SMTP) is blocked by several large cable and DSL Internet providers as well as large hosting services.

3.3 Measurement Bias

We obtain unbiased measurements from a non-trivial portion of the Internet ($\approx 31k$ BGP prefixes, cf. Table 2). However, our methodology cannot obtain measurements from networks which use content filtering to disallow Gnutella (the RSP listens on the non-default port 30494 to avoid port filtering). Thus, any extrapolation of our results to a characterization of the larger Internet is potentially biased. Networks that we have yet to measure could block more or fewer ports or different ports than those seen in existing results.

Since we wish to measure service provider discriminatory blocking, we analyze our data on the basis of BGP prefix aggregates. We reason, but do not prove, that while an individual customer of an ISP, say a corporation or university, may block Gnutella, it is unlikely that of the ISP's customers ISP block Gnutella. A single reachable node facilitates inference for that ISP. The breadth and scope of the BGP prefixes for which we have data suggest that the qualitative characteristics of blocking in our sample is likely representative of a significant fraction of the Internet. Our ongoing work seeks to further substantiate this characterization.

4 Discussion, Future Research and Conclusion

Understanding common operational practices on the Internet is particularly important as these practices are under close scrutiny in the network neutrality debates. While our data cannot answer which practices should be acceptable, the distribution of practices across different types of providers (c.f. academic and commercial) may provide insights into provider intentions.

For instance, the MIT network drops traffic destined for TCP ports 135 and 137-139, ports associated with Microsoft file sharing. With the same intent, but slightly different effect, Comcast residential broadband blocks the entire 135-139 port range [11]. Interestingly, Comcast's policy results in the *collateral blocking* of port 136, assigned to the innocuous Profile naming service [2]. The fact that MIT and other non-profit organizations block the Windows file sharing ports potentially provides justifiable evidence that Comcast's intentions in blocking the same ports are not abuses of market power. Indeed, here the motivation for blocking is based upon operators' concerns for end-user security and privacy.

Given the infancy of our scheme and the broader evolution of network neutrality, we expect this work to pose as many questions as it answers. By continuing to collect data, we can form a more complete picture of blocking, not only in terms of ports but also networks, autonomous systems and addresses.

Beyond the methodology in this paper there are several interesting and hard data analysis problems we plan to investigate. First, port-specific traceroutes to clients in our study could reveal ingress properties, filtering asymmetry and yield useful path information. By finding partially coincident AS paths with opposite blocking policies, we can infer where in the network blocking occurs. Finally, our data can shed light on the evolution of blocking over time.

Our results represent some of the first measurements in the space of neutrality and discrimination. We hope our findings will better inform the network neutrality debate by providing data on which to make informed decisions.

Acknowledgments

We thank David Clark, Neil Gershenfeld, Sachin Katti, Enoch Peserico, Karen Sollins and our reviewers for support, discussions and invaluable feedback.

References

1. kc Claffy: Top problems of the Internet and what can be done to help. In: AusCERT. (2005)
2. IANA: Well-known port numbers (2006) `http://www.iana.org/assignments/port-numbers`.

3. Clark, D.: Name, addresses, ports, and routes. RFC 814 (1982)
4. Wu, T.: Network neutrality, broadband discrimination. Telecommunications and High Technology Law **2** (2005)
5. Schewick, B.V.: Towards an economic framework for network neutrality regulation. In: Proceedings of the Telecommunications Policy Research Conference. (2005)
6. FCC: In the Matter of Madison River Communications Companies (2005) File No. EB-05-IH-0110.
7. Cerf, V.: U.S. Senate Committee on Commerce, Science, and Transportation Hearing on Network Neutrality (2006)
8. CERT: Advisory CA-2003-04 MS-SQL Worm (2003) http://www.cert.org/advisories/CA-2003-04.html.
9. Ballani, H., Chawathe, Y., Ratnasamy, S., Roscoe, T., Shenker, S.: Off by default! In: Proc. 4th ACM Workshop on Hot Topics in Networks (Hotnets-IV). (2005)
10. Masiello, E.: Service identification in TCP/IP: Well-Known versus random port numbers. Master's thesis, MIT (2005)
11. Comcast: Terms of service (2006) http://www.comcast.net/terms/use.jsp.
12. America On-Line: AOL Port 25 FAQ (2006) http://postmaster.aol.com/faq/port25faq.html.
13. Schmidt, J.E.: Dynamic port 25 blocking to control spam zombies. In: Third Conference on Email and Anti-Spam. (2006)
14. Beverly, R., Bauer, S.: The spoofer project: Inferring the extent of source address filtering on the Internet. In: Proceedings of USENIX SRUTI Workshop. (2005)
15. Mahdavi, J., Paxson, V.: IPPM Metrics for Measuring Connectivity. RFC 2678 (Proposed Standard) (1999)
16. Yang, B., Garcia-Molina, H.: Designing a super-peer network. IEEE Conference on Data Engineering (2003)
17. Ripeanu, M., Foster, I., Iamnitchi, A.: Mapping the gnutella network. IEEE Internet Computing Journal **6**(1) (2002)
18. Slyck: Slyck's P2P Network Stats (2006) http://www.slyck.com/stats.php.
19. Meyer, D.: University of Oregon RouteViews (2006) http://www.routeviews.org.

Appendix A: Ports of Interest

Port	Description
4662, 6346, 1214	Popular Peer-to-Peer
6881-6889	BitTorrent
25, 110, 143, 993	Email
27015, 27660, 7777, 7778, 28910	Popular Games
5060	Skype
2233, 500, 1494, 259, 5631	Popular VPN
80, 8080, 443	HTTP
194, 1503, 1720, 5190	Chat
20-23	Popular User Applications
53, 111, 119, 161, 179, 3306	Popular Server Applications
136	Collateral Damage
1434, 4444	Worms

Performance Limitations of ADSL Users:
A Case Study[*]

Matti Siekkinen[1,**], Denis Collange[2], Guillaume Urvoy-Keller[3],
and Ernst W. Biersack[3]

[1] University of Oslo, Dept. of Informatics, Postbox 1080 Blindern, 0316 Oslo, Norway
siekkine@ifi.uio.no
[2] France Télécom R&D, 905, rue Albert Einstein, 06921 Sophia-Antipolis, France
denis.collange@orange-ftgroup.com
[3] Institut Eurecom, 2229, route des crêtes, 06904 Sophia-Antipolis, France
{urvoy,erbi}@eurecom.fr

Abstract. We report results from the analysis of a 24-hour packet trace
containing TCP traffic of approximately 1300 residential ADSL clients.
Some of our observations confirm earlier studies: the major fraction of
the total traffic originates from P2P applications and small fractions of
connections and clients are responsible for the vast majority of the traffic.
However, our main contribution is a throughput performance analysis of
the clients. We observe suprisingly low utilizations of upload and download capacity for most of the clients. Furthermore, by using our TCP root
cause analysis tool, we obtain a striking result: in over 90% of the cases,
the low utilization is mostly due to the (P2P) applications clients use,
which limit the transmission rate and not due to network congestion, for
instance. P2P applications typically impose upload rate limits to avoid
uplink saturation that hurt download performance. Our analysis shows
that these rate limits are very low and, as a consequence, the aggregate
download rates for these applications are low.

1 Introduction

We analyze a large packet trace of clients connected to the Internet via ADSL
to investigate the causes of throughput limitations experienced by the end users.
For this purpose we use a TCP root cause analysis tool that we apply to TCP
connections. We consider throughput as the performance metric. The cause that
limits the performance of a particular connection can be located either at the
edge (sender or receiver) of a connection or inside the network. Limitations at
edge comprise the application not providing data fast enough to the TCP sender
or the TCP receiver window being too small. A network limitation may be due
to the presence of a bottleneck anywhere along the end-to-end path. We perform
root cause analysis of performance both at connection level and at client level.
Based on a packet level trace that captures the activity of over one thousand
ADSL clients during 24 hours we see that

---

[*] This work has been partly supported by France Telecom, project CRE-46126878.
[**] Work mostly done while at Institut Eurecom.

S. Uhlig, K. Papagiannaki, and O. Bonaventure (Eds.): PAM 2007, LNCS 4427, pp. 145–154, 2007.
© Springer-Verlag Berlin Heidelberg 2007

- The distribution of the client activity in terms of volume and duration is highly skewed. Most clients are active only during a short period of time. Also, most clients generate a limited amount of traffic in the order of several tens of MB, while a small number of (heavy hitter) clients upload and download hundreds of MB each.
- The utilization of the uplink and downlink is very low for most of the clients. Even heavy hitters are far from saturating their access link.
- The low utilization is mainly due to the applications that limit their rate of transfer, which is now very common for P2P applications such as eDonkey.

2   Dataset

We collected one full day (Friday March 10, 2006) of traffic generated by approximately 3000 ADSL users identified by IP addresses. We captured all IP, TCP and UDP headers of packets without any sampling or loss. The data collected on this day represents approximately 290 GB of TCP traffic in total, out of which 64% is downstream and 36% upstream. This day can be considered as a typical day in terms of volumes uploaded and downloaded by clients. Out of those 3000 clients, 1335 generated enough data to enable any root cause analysis. We consider only those clients in further analysis. In addition to the packet trace, we have a list of IP addresses that belong to local clients, which allows us to distinguish the upstream traffic from the downstream traffic. However, we do not know the clients subscription rates, i.e., their uplink and downlink capacities. The offered subscriptions were (down/up): 128/64, 512/128, 1024/256, 1024/128, 2048/128, 2048/256, 3072-4096/160, 4096-5120/192, 5120-6144/224, 6144-8640/256, and 18500/840.

We first analyzed the overall characteristics of the trace. Due to space constraints, we report here only the main findings from this study. For further details, we refer the reader to our technical report[1] which is an extended version of this paper.

The average volume of data uploaded is quite constant during the whole day, around 2GB per 30 minute period. The volume of downloaded data is less constant, around 3 GB per 30 minute period from midnight to 6 am and around 4 GB per 30 minute period for the rest of the day.

Only 5 applications generated more than 5% of the total amount of bytes: E-donkey, applications using port 80/8080, BitTorrent, email and telnet (due to a couple of hosts that generated large amount of traffic using telnet for some unknown reasons). We identify applications using port numbers and associated the TCP port range 4660-4669 to eDonkey, the ports 6880-6889 and 6969 (tracker) to BitTorrent. We do not want to declare the traffic seen on ports 80 and 8080 as Web traffic since it is likely to include also P2P traffic. The dominant category of traffic, however, is the traffic from unidentified applications, referred to as "other" traffic in the rest of the paper. Since much of today's traffic is not using fixed ports but "hiding" [2], we are not able with our port-based method to classify much of the traffic seen. Therefore, the "other" traffic represents about 50% of the total traffic. However, our root cause analysis (see next section) does not rely on the identification of the application to infer the causes for throughput limitation.

Distributions of the traffic per connection and per client are heavily skewed. Consequently, clients can be classified into two classes: heavy hitters and non heavy hitters. We identified heavy hitters as the 15% of clients that generated 85-90% of the bytes both upstream and downstream. They represent 200 clients. Those results are in line with the ones of a recent study performed on a much larger scale for Japan's residential user traffic [3]. The average amount of bytes uploaded and downloaded by a heavy-hitter client is approximately 470 MB and 760 MB, respectively, while for a non heavy-hitter those average values are 9 MB and 27 MB.

Heavy hitters also differ from non heavy hitters in terms of the set of applications they use. Overall, heavy hitters tend to use P2P applications more extensively, which is visible when looking at the identified applications (heavy hitters heavily use eDonkey) and also when merely looking at the volumes uploaded and downloaded (see above), which are significantly more symmetric for a heavy hitter than for a non heavy hitter.

Access link utilizations[1]*, uplink and downlink, are in general very low.* We observed that 80% of the clients have a downlink utilization of less than 20% and uplink utilization of less than 40% for a given 30 minute period.

Having seen that most clients achieve very low link utilization, we will now set out to investigate the causes. For this purpose, we will use some techniques referred to as root cause analysis (RCA) that has been originally proposed by Zhang et al. [4] and further refined by Siekkinen et al. [5].

3   Performance Analysis of Clients

3.1   Connection-Level Root Cause Analysis

To apply RCA, we need TCP connections that carry at least 130 data packets, which is equivalent to about 190 kB of data, if we assume MSS to be 1450 B. As pointed out in Section 2, most connections are quite small, but most of the bytes are carried in a tiny fraction of the largest connections. As a consequence, our RCA will only be able to analyze the 1% of the largest connections, which however carry more than 85% of total bytes.

We classify in a first step the packets of a connection into two groups. Each packet is either part of an **application limited period** (ALP) or a **bulk data transfer period** (BTP). Roughly speaking, the throughput of packets that are part of an ALP is *limited by the behavior of the application*. For example, an IP telephony application that produces packets at a fixed rate clearly determines (and limits) the throughput achieved. Therefore, the packets of the TCP connection carrying these data should all be put into an ALP.

---

[1] Due to lack of knowledge about clients' access link capacities, we estimated a lower bound for the capacity and, thus, obtain an upper bound for the utilization. Details are presented in the extended version[1].

The packets that are not part of an ALP will be part of a BTP. For the details on how packets get classified into ALPs and BTPs, we refer to our technical report [6].

For packets that are part of a BTP, there can be a number of causes that limit the throughput achieved, such as:

- **Network limitation:** A bottleneck limits the observed throughput. We distinguish between two types of network limitation. One is called **un-shared bottleneck** and corresponds to the case where a single connection uses the full capacity of the bottleneck link. The other type, called **shared bottleneck**, occurs when several connections share the capacity of the bottleneck link.
- **TCP end-point limitation:** The advertised receiver window is too small as compared to the bandwidth-delay product of the path, which prevents the sender to achieve a higher throughput. Note that in practice, the sender buffer size is rarely too small[5]. We count into this category also *transport limitation* which relates to the time spent for TCP ramp up[4].

The choice of the most likely limitation is based on a set of metrics computed from the packet header trace of the connection and a threshold-based classification scheme. For details, see [5] and esp. Chapter 7 of [7].

3.2   Client-Level Root Cause Analysis

We are interested in doing RCA not only at connection level but also at *client* level. We identify four types of limitations for clients, which are: (i) Applications, (ii) Access link saturation, (iii) Network limitation due to a distant bottleneck, and (iv) TCP end-point limitation. Our analysis showed the TCP end-point limitation category (described above) to be marginal in our data set. Hence, we exclude this limitation category from further discussions.

In this analysis, we focus on *active* clients. We define a client to be *active* during a period of 30 minutes if it transferred at least 100 kB during that period. For each active client we consider all the bytes transferred by all the connections of the client within a given 30-minute period. We then associate these bytes into the three considered client-level limitations. To do this association, we use the connection-level RCA as follows: All the bytes carried by the ALPs of all the connections of the client are associated to application limitation. All the bytes carried by all the BTPs that are labeled network limited (unshared or shared bottleneck) by connection-level RCA and during which the utilization is above 90% of the maximum are associated to access link saturation. All the bytes carried by the rest of the network limited BTPs during which the utilization is below 90% of the maximum are associated to network limitation due to a distant bottleneck. All the rest of the bytes transferred by the client, and not covered by these three limitations, are associated to "other" (unknown) client limitation. The amount of bytes associated with a limitation serves as a quantitative metric of the degree of that limitation for a given client during a 30-minute period.

We know from our previous work on RCA that for a single, possibly very long connection, the limitation cause may vary over time. Also, a single client may run one or more applications that will originate multiple connections. Assigning a single limitation cause to each client is therefore tricky. For this reason, we distinguish for each client between "main limitation" and "limitations experienced". As **main limitation**, we mean the limitation that affects the most number of bytes for this client. This classification is exclusive. i.e. each client belongs to a single limitation category.

On the other hand, under **limitations experienced** a single client will be considered in all the categories whose limitation causes it has experienced. Therefore, this classification is not exclusive. The results are presented in Table 1 for two 30-minute periods of the day: 4-4:30am and 3-3:30pm, which are representative for the different periods of the day. During the night time, heavy hitters dominate (70 out of 77 active uploading clients and 61 out of 83 active downloading clients), which is not surprising if one considers that heavy hitters heavily use P2P applications and P2P file transfer that can run for several hours [8]. If we look at the absolute number of clients, we see that only a small fraction of 1335 clients is active in either 30-minute period. We show only the results for the upstream direction, the ones for the downstream direction being very similar and are given in our technical report [6].

Table 1. Number of active clients limited by different causes

Upstream							
limitation cause			Total active #	application	access link	other link	other cause
main limitation	all clients	4am	77	95%	0%	4%	1%
		3pm	205	86%	6%	4%	4%
	heavy hitters	4am	70	94%	0%	4%	2%
		3pm	111	92%	2%	3%	3%
limitations experienced	all clients	4am	77	100%	0%	60%	–
		3pm	205	100%	7%	39%	–
	heavy hitters	4am	70	90%	0%	66%	–
		3pm	111	92%	5%	64%	–

Main Limitation. If we look at the main limitation cause experienced by the clients, we see that *almost all clients see their throughput performance mainly limited by the application.* This holds irrespective of the direction of the stream (upstream or downstream), of the type of client, average client or heavy hitter, and of the period of the day.

The clients that are not application limited see their throughput either limited by the capacity of the access link or the capacity of another link along the end-to-end path. Capacity limitations occur more frequently during the daytime than at night. The very limited number of cases where we observe a saturation of the access link complies with the low access link utilization observed in the preliminary analysis (Section 2).

Limitations Experienced. Besides the main limitation, we also consider *all the limitation causes* experienced by a single client. The most striking result is the difference between main limitation and limitations experienced for the "other link" limitation. As we have seen, this limitation is rarely the main limitation, while the percentage of clients that experience such limitation is between 40% and 60%, which means that while approximately half of the clients experience such network limitation, this limitation cause is not dominant. Moreover, we checked that for a given client, the amount of bytes transferred while limited by the network is generally clearly less than the amount of bytes transferred while limited by the dominant cause, i.e. the application in almost all of the cases.

3.3   Throughput Limitations Causes Experienced by Major Applications

Having done the root cause analysis on a per-client basis, we now perform application-level RCA, i.e. we investigate what are the most important applications that experience the different limitation causes, namely (i) application limited, (ii) saturated access link, and (iii) bottleneck at distant link. For each 30-minute period, we associate bytes flagged with limitations by client-level RCA to different applications based on the used TCP ports (as in Section 2).

Figure 1(a) shows the main applications that generate traffic that is application limited. If we look at the evolution of the total volume of traffic that is application limited we see very little variation in time and an upload volume almost as big as the download volume, both being around 2 GB per 30 minutes. The largest single application that generates application limited traffic is, as expected, eDonkey. However, if we look by volume, the largest category is "other", i.e. the one where we were not able to identify the application generating the traffic. The overall symmetry of upload and download volumes for the "other" category as well as a manual analysis of the traffic of some heavy hitters strongly suggest that the "other" category contains of a significant fraction of P2P traffic.

Figure 1(b) shows the main applications that saturate the access link. For this cause, no traffic originating from recognized P2P applications was seen. Instead, a significant portion of traffic saturating the uplink is e-mail. For the downlink it is mainly traffic on ports 80 and 8080 and traffic for which the application could not be identified. The fact that the traffic using ports 80 and 8080 primarily saturates only downlink suggests that it could be real Web traffic that consists of small upstream requests and larger downstream replies from the server, as opposed to P2P traffic which is typically more symmetric. If we look at the absolute volumes, we see that most of the activity is concentrated to day time, with the peak being in the early afternoon.

Figure 1(c) shows the main applications that see their throughput limited by a link that is not the client's access link. The category of other applications is clearly dominating in terms of volume. Otherwise, we observe a mixture of applications. It is expected that the set of applications is diverse since this type of network limitation can occur at any point of the network regardless of the application behavior at the particular client experiencing that limitation.

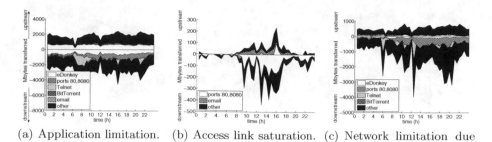

(a) Application limitation.   (b) Access link saturation.   (c) Network limitation due
to a distant link.

Fig. 1. Amount of bytes experiencing a particular root cause. Note the different scales.

In the download direction, the total traffic that is limited by a distant bottleneck reaches in the late afternoon a proportion that, in terms of volume, is almost as important as the download traffic that is application limited. The fact that this traffic peaks late afternoon[2] may be an indication of higher overall network utilization just after working hours, not only within the France Telecom network but in wider scale, that causes more cross traffic in aggregating links. Note that at the same time, the amount of traffic limited by the access link is very low (Figure 1(b)), which could indicate that these two groups represent different types of clients.

Finally, we would like to point out that a comparison of the absolute traffic volumes of Figures 1(a) – 1(c) reveal that the application limitation category represents the vast majority of the total number of transmitted bytes.

3.4   Impact of the Root Causes on Access Link Utilization

Now, we want to know how the three main root causes of throughput impact the access link utilization of the clients. We focus on link utilization and not on absolute throughput, because clients have different link capacities and we want to understand how far we are from the maximum, i.e. access link saturation.

As before, we included in the analysis for each client only the traffic of the 30-minute period for which that client achieved its highest instantaneous throughput. We computed client's link utilization during ALPs and BTPs limited by different causes. In this way, we can quantify the impact of different limitation causes on the performance. Figure 2 shows CDF plots of the results.

We focus first on uplink utilization: We see that for the case of an unshared bottleneck, the utilization is in approximately 70% of the cases very close to one, which means that in these cases the uplink of the client is the bottleneck. In the remaining 30% of cases where we observe an unshared bottleneck, we see a link utilization between 0.4 and 0.85 that can be due to a distant access downlink,

---

[2] An analysis of the IP addresses using Maxmind (http://www.maxmind.com/) revealed that most of the local clients exchange data primarily with peers/servers located in France or surrounding countries.

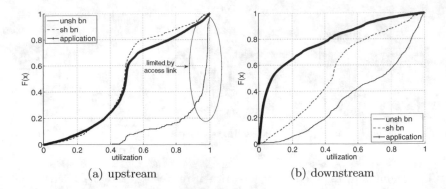

(a) upstream   (b) downstream

Fig. 2. CDF plot of access link utilization for the different root causes. For each client, we consider only traffic of the 30 min period during which that client achieved the highest instantaneous throughput of the day.

e.g. a peer that has lower downlink capacity than the uplink capacity of the local peer, or due to simply misclassification. For the two other root causes, application limitation and shared bottleneck, the clients achieve in about 60% of the cases a link utilization of less than half the uplink capacity.

If we look at the utilization of the downlink, we see that application limited traffic results most of the time in a very poor downlink utilization. Given that most of the application limited traffic is eDonkey traffic (cf. Figure 1(a)), one might be tempted to explain this low utilization by that fact that most likely the peer that sources the data has an asymmetric connection with the uplink capacity being much lower than the downlink capacity of the receiving peer[3]. However, a downloading peer has usually multiple parallel download connections, which in aggregation should be able to fully utilize the downlink capacity. The fact that this is not the case seems to indicate that many users of eDonkey use the possibility offered by todays P2P applications to limit their upload rate. Figure 3, which plots the maximum instantaneous aggregate download rates achieved per-client for different applications, further supports this hypothesis. We see that the maximum aggregate download rates of P2P applications, eDonkey and BitTorrent, are clearly below the maximum download rates of FTP and port 80/8080 traffic.

A recent study of eDonkey transfers by ADSL clients [8] found that the average file download speed achieved was only a few kB/sec. Our findings seem to indicate that such a poor performance is not due to network or access link saturation but rather due to *eDonkey users drastically limiting the upload rate of their application*.

3.5   Comparison with Other RCA Work

In [4], Zhang et al. performed flow-level root cause analysis of TCP throughput using a tool called T-RAT. They analyzed packet traces collected at high speed

---

[3] Maxmind also reported that a clear majority of the distant IPs that the heavy-hitters communicated with were clients of ISPs providing residential services.

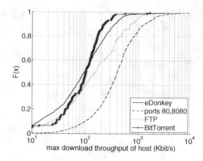

Fig. 3. CDF plot of maximum aggregate per-host download throughput computed over five second intervals

access links connecting two sites to the Internet, at a peering link between two Tier 1 providers, and at two sites on a backbone network. As results, the authors reported that, in terms of traffic volume affected, congestion (similar to network limitation in our vocabulary) was the most common limiting factor followed by host window limitation (TCP end-point in our vocabulary). It is important to notice that the data used were collected in 2001-2002. At that time, the popularity of P2P applications such as eDonkey was far from what it is today.

In order to understand whether or not our results are specific to this particular access network, we applied our RCA tool also to other publicly available packet traces collected at an ADSL access network in Netherlands (http://m2c-a.cs.utwente.nl/repository/). We looked at two 15-minute traces: one captured in 2002 and another one in 2004. A port based study similar to the one in Section 2 showed that in the 2002 trace, the applications generating most traffic were FTP and applications using ports 80 and 8080, while eDonkey and BitTorrent were dominating in the 2004 trace. We were unable to perform similar client-level study due to lack of knowledge about local client IP addresses and limited capture durations. However, simple connection-level RCA revealed that in the 2002 trace around 40% of bytes were throughput limited by the application, while this percentage was already roughly 65% in the 2004 trace, which demonstrates the impact of the increase in P2P application traffic.

4   Conclusions

We analysed one day ADSL traffic generated by more than one thousand clients. We observed that most of the clients never utilize more than a very small fraction of their upload and download capacity. TCP root cause analysis revealed that most of the user traffic is in fact *application* limited, which means that the users of P2P applications impose upload rate limits that are chosen to be very low. Other root causes that were typically observed in previous studies [4] play only a minor role: We saw some occurrences of network limitation, as well as rare occurrencies of limitations by TCP configuration issues such as too small a receiver window, but the amount of bytes transferred and affected by these causes were very small in comparison.

By severely limiting the aggregate upload rate of their P2P applications, clients certainly make sure that their P2P traffic does not interfere with concurrent activities such as Web serving or IP telephony. However, this comes at the price of very long download times, which makes the current rate limitation strategies used by P2P clients very inefficient from a users point of view.

The implication of such a low access link utilization is naturally low utilization of the entire access network, which is beneficial for the service provider. However, the utilization and traffic volumes can change dramatically in case a new type of popular P2P application is deployed or an already existing one is upgraded to utilize the uplink in a different, more effective way.

References

1. Siekkinen, M., Collange, D., Urvoy-Keller, G., Biersack, E.W.: Application-level performance of ADSL users. Technical report (2006) http://www.ifi.uio.no/~siekkine/pub.html.
2. Karagiannis, T., et al.: Is P2P dying or just hiding? In: Proc. Globecom 200. (2004)
3. Cho, K., Fukuda, K., Esaki, H., Kato, A.: The impact and implications of the growth in residential user-to-user traffic. In: Proceedings of ACM SIGCOMM '06, New York, NY, USA, ACM Press (2006) 207–218
4. Zhang, Y., Breslau, L., Paxson, V., Shenker, S.: On the characteristics and origins of internet flow rates. In: Proceedings of ACM SIGCOMM '02, Pittsburgh, PA, USA (2002)
5. Siekkinen, M., Urvoy-Keller, G., Biersack, E., En-Najjary, T.: Root cause analysis for long-lived tcp connections. In: Proceedings of CoNEXT. (2005)
6. Siekkinen, M., Urvoy-Keller, G., Biersack, E.W.: On the impact of applications on tcp transfers. Technical report, Institut Eurecom (2005) http://www.eurecom.fr/~siekkine/pub/RR-05-147.pdf.
7. Siekkinen, M.: Root Cause Analysis of TCP Throughput: Methodology, Techniques, and Applications. PhD thesis, Institut Eurécom / Université de Nice-Sophia Antipolis, Sophia Antipolis, France (http://www.ifi.uio.no/~siekkine/pub.html) (2006)
8. Plissonneau, L., Costeux, J.L., Brown, P.: Detailed analysis of edonkey transfers on adsl. In: Proceedings of EuroNGI. (2006)

Fast Classification and Estimation of Internet Traffic Flows

Sumantra R. Kundu, Sourav Pal, Kalyan Basu, and Sajal K. Das

Center for Research in Wireless, Mobility and Networking (CReWMaN)
The University of Texas at Arlington, TX 76019-0015
{kundu,spal,basu,das}@cse.uta.edu

Abstract. This paper makes two contributions: (i) it presents a scheme for classifying and identifying Internet traffic flows which carry a large number of packets (or bytes) and are persistent in nature (also known as the *elephants*), from flows which carry a small number of packets (or bytes) and die out fast (commonly referred to as the *mice*), and (ii) illustrates how non-parametric Parzen window technique can be used to construct the probability density function (pdf) of the elephants present in the original traffic stream. We validate our approach using a 15-minute trace containing around 23 million packets from NLANR.

1 Introduction

There are two main aspects to the problem of Internet traffic flow characterization: (i) *how* to efficiently collect the flows, and (ii) how to accurately *infer* overall traffic behavior from the collected data. Due to limitations in hardware capabilities, it has been illustrated in [7] [22] how exhaustively collecting all packets of a flow does not scale well at high link speeds (OC-48+). Thus, current approaches to flow characterization are either based on: (i) *statistical sampling* of the packets [5][6], or (ii) *inferring* traffic characteristics primarily based on flows which carry a large number of packets (or bytes) and are long-lived in nature) (i.e., the *elephants*) while ignoring flows which carry very small number of packets (or bytes) and are short-lived in nature (i.e., the *mice*) [10], or (iii) using appropriate *estimation algorithms* on lossy data structures (e.g., bloom filters, hash tables) [7][17] for recovering lost information. However, even in sampled traffic, separation of elephants and mice is a cumbersome task [8] since there exists no standard approaches to drawing the line between the two.

In this paper, we show that it is indeed possible to provide an analytical framework for identifying and classifying packets as elephants or mice by applying Asymptotic Equipartition Property (AEP) from Information Theory [18]. It is based on the observation that *all mice die young and are large in number*; while the proportion of elephants *is small in number* (around $1\% - 2\%$ of the traffic volume) and they have average longevity varying from a few minutes to days [1]. If the state space of the Internet flows is visualized to be an ergodic random process, then the existence of *typical sequence*, as defined by AEP, identifies the presence of elephants in the traffic volume. Such an approach requires

S. Uhlig, K. Papagiannaki, and O. Bonaventure (Eds.): PAM 2007, LNCS 4427, pp. 155–164, 2007.

no prior knowledge of flow distribution, does not suffer from the side effects of false positives associated with Bayesian analysis, and involves minimal packet processing. We compare our approach with the well-known method of identifying packets as elephants based on the average flow longevity of greater than 15 minutes [8]. Our results from initial analysis on a single 15-minute traffic trace from NLANR [15] indicates that there exists a possibility that using definite values of longevity as cutoff limits for classifying flows as elephnats might overestimate the frequency of occurence of such flows. In the second part of the paper, we use a statistical non-parametric estimation technique based on the Gaussian kernel function for accurately estimating the density function of the underlying traffic, considering the probability density function (pdf) of only the elephants.

The remainder of the paper is organized as follows. In Section 2, we present the theory and online framework for classifying traffic flows into elephants and mice. This is followed by Section 3 which briefly presents the theory for estimating the distribution of the elephants. Evaluating the effectiveness of our approach is carried out in Section 4 with conclusions in Section 5.

2 An Online Framework for Identifying the Elephants

In this work, we define traffic *flows* to refer to packets with similar attributes. For example, a flow might be defined to consist of packets having identical values of five-tuple (source address, destination address, source port, destination port, protocol) or might be defined to comprise of packets matching specific payload information (e.g., group of all TCP packets with payload containing the string "crewman"). Thus, flows can be characterized by packet headers, payloads or a combination of both. The *size* of a flow is the number of packets (or bytes) belonging to the flow and the *duration* of a flow is its lifetime. Let $[\mathcal{F}] = \{F_1, F_2, \ldots, F_i, \ldots, F_N\}$ be a sequence of N FlowIDs $\{1, 2, \ldots i, \ldots N\}$, where each FlowID, F_i, is an index (i.e., a number between 1 and N) used to identify each flow in the underlying traffic. Denote $|F_i|$ to represent the number of packets belonging to the flow with FlowID F_i. It is important to note that the sequence $[\mathcal{F}]$ is sorted by increasing cardinality of the number of packets present in each FlowID. Under such circumstances, the flow classification problem is to identify and separate the F_is that define the elephants and the mice. Now let us now consider an ergodic and discrete random process where each F_i is an independent variable drawn from the state space of $[\mathcal{F}]$. The state space of $[\mathcal{F}]$ consists of all possible FlowIDs. However, the random variables are not identically distributed. Denote $\{f_i\}$ to be the set of possible outcomes of F_i with $f \in [\mathcal{F}]$. Let us represent the probability mass function (pmf) of the sequence $\{F_i\}_{i=1}^N$ by: $P(F_1 = f_1, \ldots, F_N = f_N) = p(f_1, \ldots f_N)$ Let $H(\mathcal{F}) = H(F_1, F_2, \ldots, F_N)$ denote the *joint entropy* of the sequence $\{F_i\}_{i=1}^N$ and denote $\bar{H}_{\mathcal{F}}$ to be the *entropy rate* of $\{F_i\}_{i=1}^N$. Then, $H(\mathcal{F})$ and $\bar{H}_{\mathcal{F}}$ are defined as follows [18]:

$$H(\mathcal{F}) = H(F_1, F_2, \ldots, F_N) = \sum_{i=1}^{N} H(F_i | F_{i-1}, \ldots F_1) \tag{1}$$

$$\bar{H}_{\mathcal{F}} = \frac{1}{N} H(\mathcal{F}) \tag{2}$$

Since according to our assumption, the F_is are independent, Equation (1) reduces to: $H(\mathcal{F}) = \sum_{i=1}^{N} H(F_i)$ which is the summation of the individual entropies of the flow. At this point, it is worth mentioning that it is possible to *estimate* $H(\mathcal{F})$ without considering individual flow entropies [13]. However, this is not considered in this work.

Definition 1. *The set of elephants present in a sampled traffic is represented by the sequence, $\{F_1 F_2 \ldots F_{N'}\}$, where $N' \ll N$ denotes the total number of elephants.*

Definition 1 provides us with the set of all packets which belong to the set of elephants. Since our aim is to identify the sequence $\{F_1, F_2, \ldots, F_{N'}\}$, we need to isolate the sequence of FlowIDs that form the high probability set. If we visualize the set, $[\mathcal{F}]$, as an information source, then the existence of the above sequence of FlowIDs is governed by the probability of occurrence of a jointly typical sequence based on AEP. Note that the results based on AEP hold true only when the number of FlowIDs present in the sampled traffic volume is very large. Now considering the fact that there can be several sets of typical sequences, we have the following lemma for the set of elephants:

Lemma 1. *For traffic volumes with large number of FlowIDs (i.e., $N \to \infty$), the occurrence of the sequence $\{F_1, F_2, \ldots, F_{N'}\}$, $N' \ll N$, is equiprobable and approximately equal to $2^{-N\bar{H}_{\mathcal{F}}}$.*

Lemma 1 follows directly from the property of AEP. In view of the above, we can say that out of all the possible FlowIDs, that sequence which belongs to the typical set has the maximum concentration of probability. The sequences outside the typical set are *atypical* and their probability of occurrence is extremely low. As evident from the above lemma, a typical sequence implies that FlowIDs in the typical set are associated with a large number of packets. If we consider the distribution of FlowIDs in the Internet traffic, we can easily correlate this property with the Zipf distribution of Internet flows. Hence, it is not surprising that most of the elephants belong to the typical set. However, what is the guarantee that such a sequence really exists?

Definition 2. *The joint entropy, $H(\mathcal{F})$ for a stationary, stochastic process is a decreasing sequence in N and has a limit equal to its entropy rate.*

Definition 2 implies that the probability of correctly identifying elephants *increases* with the corresponding increase in traffic volume. This observation is of fundamental nature since it enables us to *scalably* create an approximate list of LLFs (i.e., a typical sequence), while avoiding needless complex computation.

2.1   Algorithm for Flow Classification

Let $L\{\}$ be the list of empty LLFs and m the number of FlowIDs observed at the time instant the classification algorithm is being executed. Denote P_i^p and P_i^b to indicate the probability of occurrence of FlowID F_i in the sequence $F_1, \ldots, F_N$, when considering the number of packets and payload bytes, respectively. Then,

$$P_i^p = |F_i| / \sum_{i=1}^{n} |F_i| \text{ and } P_i^b = \frac{\text{(cumulative payload carried by } F_i)}{\text{(total bytes observed at time instant t)}} \tag{3}$$

The pseudo-code of the classification algorithm is as follows:

1. Initialize list $L\{\} := $ null
2. $m := $ number of FlowIDs in current context;
Loop: over all sampled $\{F_i\}$
   3. calculate probability P_i (Note: if the aim is to identify the set of elephants based on the number of packets, replace P_i with P_i^p. Similarly, for identifying the set of elephants based on payload size, replace P_i with P_i^b,), for each $\{F_i\}$ using Equation 3
   4. calculate $H(\mathcal{F})$ and $\bar{H}_{\mathcal{F}}$
   5.   if $p(F_i) \geq 2^{-n\bar{H}_{\mathcal{F}}}$
   6.   add F_i to L
Done
7. List $L\{\}$ contains the set of traffic flows which are elephants.

3   Estimating the Density of Elephants Flows

We employ the Parzen window [19] technique (explained below) on the set $\mathcal{F}$ for determining the density of the identified elephants. Note that the likelihood estimator from the coupon collector problem [23] can be employed on the sampled set of all elephants in order to identify the set of all elephants present in the underlying traffic. However, that aspect is not presented in this work. The standard method is to choose a well-defined kernel function (e.g., Gaussian) of definite width and convolve it with the known data points. Let $\hat{f}_h(x)$ be the pdf of the random variable $\mathcal{X}$ we are trying to estimate for the set $\mathcal{F}$ and be defined as [19]:

$$\hat{f}_h(x) \approx \frac{1}{Nh} \sum_{i=1}^{N} \psi\left(\frac{x - x_i}{h}\right) \tag{4}$$

where $\{x_i\}_{i=1}^{N}$ are the data points of $\mathcal{X}$ and $\psi(\cdot)$ is a suitable kernel smoothing function of width h, also referred to as the bandwidth of $\psi(\cdot)$. In this approach, the estimated pdf is a linear combination of kernel functions centered on individual x_i. In Equation (4), the bandwidth factor h is the most important term in the estimation process [20]. The optimal value of the kernel window h can

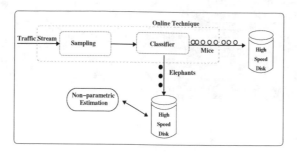

Fig. 1. On-line classification of traffic streams. The pdf of elephant flows in sampled traffic stream is estimated using the non-parametric Parzen window technique.

be calculated by minimizing the integrated mean square error (IMSE) between $f(x)$ (original pdf) and $\hat{f}_h(x)$; i.e.,

$$\text{minimize}\left\{\int \left\{\hat{f}_h(x) - f(x)\right\}^2 dx\right\}.$$

In general, the process of finding the optimal window size is cumbersome as we do not know beforehand the nature of the density function that we are trying to estimate. Since the shape (degree of smoothness) of $\hat{f}_h(x)$ is closely related to the kernel function used, we use the Gaussian kernel function to eliminate "noises" in the pdf estimation. Thus:

$$\psi(u) = \frac{1}{\sqrt{2\pi}}\exp\left(-\frac{u^2}{2}\right) \tag{5}$$

Corresponding to the Gaussian kernel, the bandwidth h can be approximated using *Silverman's rule of thumb* [21] that satisfies the IMSE criteria. Consequently, h is defined as: $h = 1.06\,\hat{\sigma}\,N^{-1/5}$ where $\hat{\sigma} = \sqrt{\frac{\sum_{i=1}^{N}(x_i-\bar{x})^2}{N}}$ denotes the standard deviation of the sample.

4   Performance Evaluation

In this section, we evaluate the performance of our algorithm using packet traces obtained from NLANR [15]. We compare our approach with the results of Mori et al. [8] in the figures) for comparing the number of elephants detected in the traffic stream. Specifically, we use three traces: (i) `20040130-133500-0.gz`, (ii) `20040130-13400-0.gz`, and (iii) `20040130-134500-0.gz`. The cumulative duration of the three files is 900 seconds and contains 23.2 million packets. They subsequently map to $618,225$ FlowIDs, where each FlowID is defined using the number of packets.

4.1   Identifying the Elephants

In Figure 2, we plot the number of elephants predicted using our classification algorithm and compare it with the approach of [8]. We have used the frequency

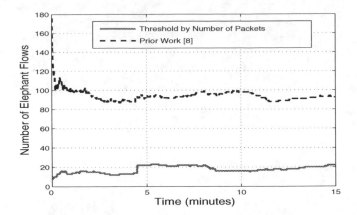

Fig. 2. Time series of the occurrence of elephants as estimated using our algorithm versus as predicted using the approach of [8]

of occurrence of the packets as the basis for calculating the flow probabilities. Apart from the already known facts that the proportion of elephants are small in number (0.0035% in our case), two important conclusions can be immediately drawn from this figure:

- the set of elephants detected during the *initial phase* (first 5 minutes for the traffic traces under consideration) of our algorithm identifies FlowIDs that exhibit *bursty behavior*. On close analysis of the traffic traces, we found that this is indeed the case and is due to the fact that such FlowIDs cause immediate concentration of the probability mass function of the entire traffic sample.
- the *proportion of elephants* classified using the frequency of occurrence of packets (i.e. probability P_i^p) is almost equal in extent to those detected by considering the volume (bytes) of traffic (i.e. probability P_i^b). Notice that, using the approach of [8], the number of elephants are estimated at around $85 - 90$. If the traffic traces beyond 5 minutes are considered (not shown in this study), the approach of [8] exhibits a *decreasing trend*.

The occurrence of mice, however, shows well-established behavior. They are large in number but grow with the continuation of traffic stream.

4.2 Traffic Distribution: Elephants and Mice

In Figures 3 and 4, we analyze the traffic carried by elephants and mice when considering the frequency of occurrence of packets and the the volume of traffic carried by each FlowID as the basis of flow probability calculation. While 99% FlowIDs carry 70% of network traffic, elephant flows (less than 1%) carry 30% of the traffic. Such dynamics is unaffected if we choose the frequency or the volume of traffic as the basic for probability calculation.

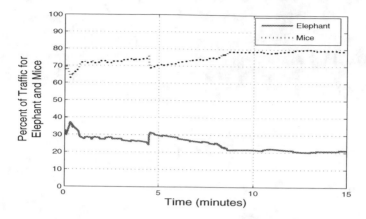

Fig. 3. Traffic Distribution of Elephant Flows considering the frequency of occurrence of packets in each flow (i.e., P_i^p)

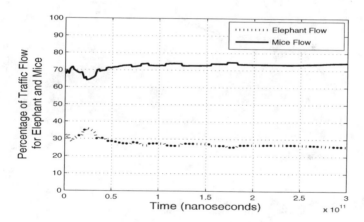

Fig. 4. Traffic Distribution of Elephant Flows considering the volume of traffic in each flow (i.e., P_i^b)

4.3 Entropy Distribution: Elephants and Mice

In Figures 5 and 6, we show the *temporal variation* of the entropy of the ratio of entropy between the elephants and mice, where only of the FlowIDs considered to be mice. During the first 500msecs of input traffic, we observe a *dip* in the entropy of the elephant flows. This is due to the presence of bursty elephant flows which causes a temporary increase in the probability of the set of elephants. However, as the experiment continues, the entropy of the mice increases while the entropy of the elephant flows decreases. Since the entropy of the typical set is a *decreasing sequence* with respect to the number of FlowIDs, the probability and proportion of FlowIDs classified as elephants increases.

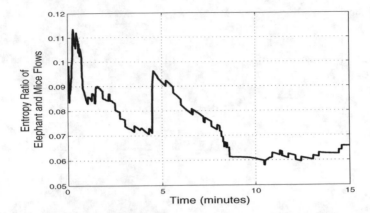

Fig. 5. Ratio of entropy between the elephants and mice

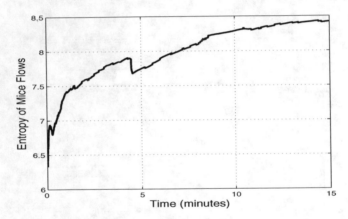

Fig. 6. Temporal distribution of Entropy of mice

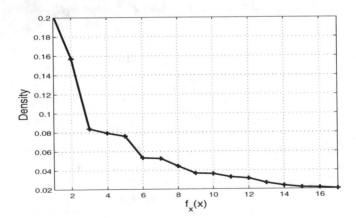

Fig. 7. Original pdf of the elephants present in the traffic stream

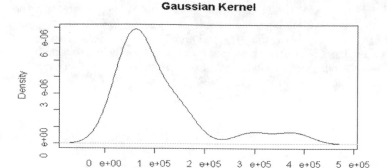

Fig. 8. Pdf estimated using parzen window technique

This unique trend of entropy variation guarantees conservative, yet accurate flow classification of high traffic volumes.

Estimating the density function of the distribution. In Figures 7 and 8, we plot the pdf of the elephants present in the original traffic stream versus the pdf of elephants identified using our approach based on the Parzen window technique. Observe that the trend observed is similar in both the cases.

5   Conclusions

In this paper we have focused on classifying and estimating the properties of elephants and mice based on AEP from Information Theory. Although considerable attention has been directed in identifying high and low traffic volumes, we feel that our approach using typical sequences simplifies the problem to a large extent and provides a standard yardstick for defining such long-lived-flows. We have evaluated our algorithm with the approach of [8], and have observed that our approach is able to identify bursty elephant flows and at the same time does not overestimate the number of occurrence of the elephants. As part of future work, we would like to carry out these observations on NLANR trace of more than one hour duration and see how the classification and estimation algorithms perform if the input traffic becomes smooth with non-negligent coefficient of variation [14].

References

1. J. S. Marron, F. Hernandez-Campos and F. D. Smith, *"Mice and Elephants Visualization of Internet Traffic"*, available online at *citeseer.ist.psu.edu/531734.html*.
2. A. Kuzmanovic and E. Knightly, *"Low-Rate TCP-Targeted Denial of Service Attacks (The Shrew vs. the Mice and Elephants)"*, *in Proceedings of ACM SIGCOMM*, August 2003.

3. Y. Joo, V. Riberio, A. Feldmann, A. C. Gilbert and W. Willinger, *"On the impact of variability on the buffer dynamics in IP networks"*, *Proc. 37th Annual Allerton Conference on Communication, Control and Computing*, September 1999.

4. K. Papagiannaki, N. Taft, S. Bhattacharya, P. Thiran, K. Salamatian and C. Diot *"On the Feasibility of Identifying Elephants in Internet Backbone Traffic"*, *Sprint ATL Technical Report TR01-ATL-110918*, November 2001.

5. N. Duffield, C. Lund and M. Thorup, "Properties and Prediction of Flow Statistics from Sampled Packet Streams", *in Proc. of ACM SIGCOMM Internet Measurement Workshop*, Nov. 2002.

6. N. Duffield, C. Lund and M. Thorup, "Estimating Flow Distributions from Sampled Flow Statistics", *in Proc. of ACM SIGMETRICS*, August 2003.

7. A. Kumar, J. Xu, O. Spatschek and L. Li, "Space-Code Bloom Filter for Efficient Per-Flow Traffic Measurement", *IEEE INFOCOM*, August 25-29, 2004.

8. T. Mori, M. Uchida, R. Kawahara, J. Pan and S. Goto, *"Identifying elephant flows through periodically sampled packets"*, *in Proceedings of the ACM SIGCOMM Workshop on Internet Measurment Workshop (IMW)*, 2004.

9. J. Sommers and P. Barford, "Self-Configuring Network Traffic Generation", *in Proc. of ACM SIGCOMM Internet Measurement Workshop*, October 25-27, 2004.

10. C. Estan and G. Varghese, "New directions in traffic measurement and accounting: Focusing on the elephants, ignoring the mice", *ACM Trans. Comput. Syst.*, vol. 21, no. 3, 2003.

11. K. Papagiannaki, N. Taft, S. Bhattacharyya, P. Thiran, K. Salamatian and C. Diot, *"A pragmatic definition of elephants in internet backbone traffic"*, *in Proceedings of the ACM SIGCOMM Workshop on Internet Measurment Workshop (IMW)*, 2002.

12. J. Wallerich, H. Dreger, A. Feldman, B. Krishnamurthy and W. Willinger, "A Methodology for Studying Persistency Aspects of Internet Flows", *in ACM SIGCOMM Computer Communication Review*, vol. 35, Issue 2, pp. 23 - 36, 2004.

13. A. Lall, V. Sekhar, M. Ogihara, J. Xu and H. Zhang, "Data Streaming Algorithms for Estimating Entropy of Network Traffic", *in Proc. of ACM SIGMETRICS*, June 2006.

14. J. Cao, W. S. Cleveland, D. Lin and D. X. Sun, "Internet Traffic Tends Towards Poisson and Independent as Load Increases",*in Nonlinear Estimation and Classification*, New York, Springer-Verlag, 2002.

15. NLANR AMP Website: http://pma.nlanr.net/Special/

16. N. Brownlee, "Understanding Internet Traffic Streams: Dragonflies and Tortoises", *IEEE Communications Magazine*, Cot. 2002.

17. A. Kumar, M. Sung, J. Xu and L. Wang, "Data Streaming Algorithms for Efficient and Accurate Estimation of Flow Size Distribution", *ACM SIGMETRICS*, August 25-29, 2003.

18. T. M. Cover and J. A. Thomas "Elements of Information Theory", John Wiley, 1991.

19. E. Parzen, "On estimation of a probability density function and mode," *Time Series Analysis Papers*. San Francisco, CA: Holden-Day, 1967.

20. S. Raudys, "On the effectiveness of Parzen window classifier," *Informatics*, vol. 2, no. 3, pp 435-454, 1991.

21. B. W. Silveman, *Density Estimation for Statistics and Data Analysis*, Chapman and Hall, 1986.

22. S. R. Kundu, B. Chakravarty, K. Basu, and S. K. Das, "FastFlow: A Framework for Accurate Characterization of Network Traffic", *IEEE ICDCS*, 2006.

23. M. Finkelstein, H. G. Tucker and J. A. Veeh, *Confidence Intervals for the number of Unseen Types*, Statistics and Probability Letters, vol. 37, pp. 423-430, 1998.

Early Recognition of Encrypted Applications

Laurent Bernaille and Renata Teixeira

Université Pierre et Marie Curie - LIP6-CNRS
Paris, France

Abstract. Most tools to recognize the application associated with network connections use well-known signatures as basis for their classification. This approach
is very effective in enterprise and campus networks to pinpoint forbidden applications (peer to peer, for instance) or security threats. However, it is easy to use
encryption to evade these mechanisms. In particular, Secure Sockets Layer (SSL)
libraries such as OpenSSL are widely available and can easily be used to encrypt
any type of traffic. In this paper, we propose a method to detect applications in
SSL encrypted connections. Our method uses only the size of the first few packets
of an SSL connection to recognize the application, which enables an early classification. We test our method on packet traces collected on two campus networks
and on manually-encrypted traces. Our results show that we are able to recognize
the application in an SSL connection with more than 85% accuracy.

1 Introduction

Accurate classification of traffic flows is an essential step for network administrators
to detect security threats or forbidden applications. This detection has to happen as
early as possible, so that administrators can take appropriate actions to block or control
the problem. Given that the simple inspection of IANA-assigned port numbers is no
longer a reliable mechanism for classifying applications [1], many campus or enterprise
networks now use content-based mechanisms. These mechanisms search the content
of packets for well-known application signatures [2,3,4]. Although very effective and
accurate, content-based mechanisms are easy to evade by using encryption. To make
matters worse, Secure Sockets Layer (SSL), which can easily be used to encrypt any
application communication, is widely available. In this paper, we design a classifier able
to detect the underlying application in encrypted SSL connections.

Before constructing a classifier for encrypted traffic, we characterize the use of SSL
on two campus networks by studying packet traces. This characterization sheds light
on the prevalence of SSL in today's networks, the SSL versions in use, and the types
of application that use encryption. We see an increase in the use of SSL, which coin-
cides with the surge of new applications that use SSL. For instance, Bittorrent clients
(Azureus and uTorrent) now offer SSL encryption as a way to hide from content-based
blocking of peer-to-peer applications. These factors indicate that SSL usage will continue to increase.

Our classifier for encrypted traffic builds on two observations. First, SSL does not
modify significantly the the number of packets, their size, and their inter-arrival time
[5,6]. Second, TCP connections can be classified based on flow-level information such

S. Uhlig, K. Papagiannaki, and O. Bonaventure (Eds.): PAM 2007, LNCS 4427, pp. 165–175, 2007.
© Springer-Verlag Berlin Heidelberg 2007

as duration, number of packets and mean inter-arrival time [7,8,9,10,11]. We extend the classifier presented in [12,13] to identify applications in encrypted SSL connections. This classifier, which we refer to as *early identification*, identifies the application associated with a TCP connection using only the first few packets in the connection. Our method to identify encrypted traffic involves two steps. First, we detect SSL traffic. Then, we apply early identification to the first encrypted application packets to recognize the underlying application. With this method, we recognize encrypted applications with more than 85% accuracy.

After comparing our traffic classification method with previous work in Section 2, we present the packet traces we used to train our classifier and to evaluate our mechanism in Section 3. Section 4, briefly introduces SSL, describes a content-based approach to identify SSL connections, and characterizes SSL usage in our traces. Section 5 presents our classification mechanism and Section 6 evaluates it. We conclude in Section 7 with a summary of our contributions and a discussion of future directions.

2   Related Work

Although any flow-level classifier [7,8,9,10,11,13,14] can potentially identify encrypted applications, this paper is the first to design an application-recognition mechanism for encrypted traffic and test it on real SSL traffic. We choose to extend the classifier presented in [12,13], because it can recognize the application associated with a TCP connection early. All of the other classifiers rely on statistics on the whole connection, which prevents them from being used online. An alternative method [15] uses connectivity patterns for each host in the network. This approach could work for encrypted traffic, but its goal is different from ours: it finds services associated to hosts, whereas we classify single TCP connections.

The methods presented in [14,16] are the only ones that share our goal of classifying encrypted traffic. As other flow-level classifiers, however, this mechanism also requires *all* packets in the connection before classifying it. In addition, we perform a measurement-based study that first characterizes the usage of SSL in two campus networks and then evaluates our mechanism against real SSL connections, whereas their classifier was only evaluated under simulated encrypted traffic.

3   Packet Traces with Encrypted Traffic

Our study relies on two sets of data: packet traces collected at the edge of two campus networks and manually-generated traces. Packet traces allow us to characterize the usage of SSL in operational networks, when manually-generated traces help us validate our classification mechanism.

We used two one-hour traces collected at the edge of the Paris 6 network, in 2004 and in 2006 (referred to as P6-2004 and P6-2006, respectively) and a packet trace collected at the edge of the UMass campus in 2005. Both traces collected on the Paris 6 network contain packet payload, which allows the identification of SSL versions and options. The UMass trace only captures 58 bytes for each packet, because of privacy and security reasons. Fortunately, many packets do not contain any options (IP or TCP) and, without options, 58 bytes capture four bytes of TCP payload, which is enough to accurately identify SSL connections and the versions they use. In the UMass trace, 50% of connections on SSL standard ports had 4 bytes of TCP payload.

To validate our method, we needed a ground truth, or SSL connections for which the underlying application is known. We use two methods to obtain this ground truth. First, we filter the Paris 6 traces to keep only connections directed to well known HTTPS and POP3S servers in the university. To extend our validation to other types of traffic, we also manually encrypted traces consisting of other applications. We replay packet traces over an encrypted tunnel and capture the resulting connections. We use three machines, say A, B and a controller. Machine A represents the server of the TCP connection and machine B the client. First, we establish a tunnel between A and B using stunnel[1]. Then, the controller parses an existing packet trace. For each packet in a TCP connection, if the packet was sent from the TCP server, the controller asks A to send the packet to B over the encrypted tunnel, otherwise it asks B to send it.

4   Analysis of SSL Traffic

This section presents a brief background on SSL and a content-based method to identify SSL connections in packet traces. We end with a characterization of SSL in our traces.

4.1   Description of SSL

Secure Sockets Layer (SSL) provides authentication and encryption to TCP connections. SSL runs between the transport layer (usually TCP) and the application layer. Three different versions of SSL have been developed: SSLv2 [17], SSLv3.0 [18] and TLS [19]. As SSL version 2 (SSLv2) presents several security flaws, its use is now strongly discouraged. Its follow-up version is SSL version 3 (SSLv3.0). The latest version, Transport Layer Security (TLS or SSLv3.1), is the standard specified by the Internet Engineering Task Force (IETF) and is similar to SSLv3.0. The differences between SSLv3.0 and TLS are minor (for instance, types of ciphers supported, pseudo-random functions and padding policies) and do not affect the handshake or the packet sizes. Therefore, we use SSLv3 to refer to both protocols.

Figure 1 presents an SSL handshake for SSLv2. The exact messages exchanged differ for SSLv3, but the main steps remain the same. First, the client and the server negotiate the SSL version they are going to use and choose an encryption algorithm. Second, the server authenticates itself to the client (the client might do likewise if required by the server) and both peers negotiate an encryption key. Finally, they terminate the handshake, and can start exchanging application data over the encrypted channel.

4.2   Identifying SSL Connections

Many applications can be detected based on a well-known signature (for instance "GET /index.html HTTP/1.1" for HTTP traffic). Unfortunately, there is no such pattern for SSL. The server Hello packet sets the configuration of the connection (in particular,

---

[1] http://www.stunnel.org

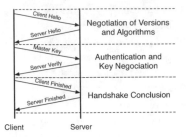

Fig. 1. Example of SSL handshake (SSLv2)

version and encryption algorithm). Therefore, we can analyze this packet to verify if a connection is using SSL and determine the version. SSL connections rely on SSL records transported using TCP. Each SSL record starts with a SSL header that is sent unencrypted, and is followed by either SSL configuration options (during the SSL handshake) or by the encrypted application payload. The first two bits in SSLv2 headers are always 1 and 0, the following 14 bits contain the size of the SSL record and the third byte is the message type (1 for "Client Hello" and 4 for "Server Hello"). The first byte of SSLv3.x (i.e. SSLv3.0 or TLS) packets is the message type (22 for configuration records and 23 for records with encrypted application payload). The second and third bytes indicate the major and minor versions (3 and 0 for SSLv3.0 or 1 for TLS).

Let $bit_i[x]$ be the content of bit x in the payload of packet i in the connection, $bit_i[x : y]$ the integer represented by the sequence of bits from x to y, $Byte_i[z]$ the value of byte z, and $Size_i$ the payload size of packet i (computed from fields in IP and TCP headers: Internet Header Length, Total Length and Data Offset). We summarize the decision process to determine if a connection is using SSL and the associated version in the following algorithm:

```
If bit₂[0] = 1 and bit₂[1] = 0 and bit₂[2 : 15] = Size₂ and Byte₂[3] = 4
    Connection is an SSLv2 connection
Else If Byte₂[1] = 22 and Byte₂[2] = 3
    If Byte₂[3] = 0
        Connection is an SSLv3.0 connection
    Else If Byte₂[3] = 1
        Connection is a TLS connection
    Else Connection is not using SSL
Else Connection is not using SSL
```

4.3 Description of SSL Traffic

We applied our identification mechanism to three traces: P6-2004, P6-2006 and UMass. Table 1 shows the proportion of the SSL version found in each trace. We see that most SSL traffic consists of SSLv3.0 and TLS, although there are still a few instances of SSLv2 in the P6-2006 trace. By comparing the proportion of SSL connections in the P6 traces from 2004 and 2006 (4.6% and 8.6%, respectively), we see a sharp increase in the usage of SSL. This trend is supported by the SSL surveys achieved by netcraft [20] (in 2005 only, the use of SSL in the web servers they surveyed increased by 30%). On the UMass trace, this proportion is lower. This difference is because the P6 traces consist only of academic traffic, whereas the UMass campus also has a dorm and, therefore, contain many other types of applications (such as online games).

An interesting observation is the proportion of non-SSL traffic in connections using standard SSL ports (labeled as "SSL Port but not SSL"). We studied this traffic in detail. In P6-2004, all non-SSL connections on SSL ports were non-encrypted traffic using port 443 (probably misconfigured web servers). In P6-2006, we still find this non-encrypted HTTP traffic, but also observe other types of traffic. For instance, the trace contains un-encrypted SIP traffic (VoIP connections from Instant Message softwares using port 443 to avoid firewalls), and HTTP connections using the CONNECT method. This method is used when a web client connects to an SSL web page using a proxy. However, the contacted servers were not proxies but web servers. It turned out these clients were try-ing to connect to SMTP servers using the web servers as TCP proxies, probably to send spam (this method works for misconfigured Apache servers with proxy capability). In the UMass trace, we also found Bittorent connections using port 443 to avoid firewalls.

Finally, using the detection method presented in Section 4.2, we evaluated the pro-portion of SSL on ports not usually associated with SSL ("SSL on non-SSL Ports"). This proportion is not negligible and is even increasing on the P6 network. This indi-cates that SSL is spreading to applications for which it was not formerly used.

Table 1. SSL versions

Trace	Total Connections	SSL Connections	SSLv2	SSLv3	TLS	SSL Port but not SSL	SSL on non-SSL Ports
P6-2004	0.5M	4.6%	0.6%	81.0%	18.4%	1.9%	1.1%
P6-2006	1.0M	8.6%	0.2%	53.2%	46.6%	1.1%	4.2%
UMass	1.7M	1.2%	0 %	48%	52%	5.0%	1.5%

5   Classification Mechanism

Our characterization in the previous section shows that SSL traffic is increasing. We now present a methodology to identify the applications in SSL connections. Figure 2 describes our classification mechanism. This classifier takes as input a stream of packets from a TCP connection and outputs the application associated to the connection. It runs in three steps: recognition of SSL connections, detection of the first packet containing application data, and recognition of the encrypted applications.

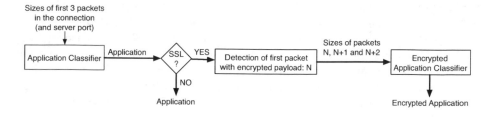

Fig. 2. Classifier Overview

5.1 Recognition of SSL Traffic

We use early classification [13] to recognize SSL traffic. We construct this classifier in two phases: training and online classification. The training phase applies a clustering algorithm to a set of sample TCP connections, which we call the training data. We represent each connection in this data set in a P-dimensional space using the sizes and directions of the first P data packet ([13] presents a detailed analysis that justifies using packet sizes instead of other features such as Inter-Arrival Time). Resulting clusters contain one or more applications. The online classification uses a heuristic to assign a TCP connection to one of the clusters and another heuristic to label it with one of the applications in the cluster.

For this study, we use a training data set composed of the following applications: Http, Ftp, Nntp, Pop3, Smtp, Ssh, Msn, Bittorent, Edonkey, SSLv2 and SSLv3. We apply a signature-based filtering method on the P6-2004 trace to select 500 random connections for Http, Ftp, Nntp, Pop3, Smtp, Ssh, Msn and Edonkey. Since the amount of identifiable Bittorent traffic in our traces is small, we manually generate a Bittorent trace from which we select 500 connections. Additionally, we use the method presented in Section 4 to select 500 SSLv2 and SSLv3 connections from the P6-2006 trace.

We applied our clustering mechanism to this training set. We use a clustering algorithm based on Gaussian Mixture Model [13]. We find that using the first three packets and 35 clusters gives good results (the method to choose the number of packets and clusters is described in [13]). To assign a new connection to a cluster, we compute the probability that this connection belongs to each cluster and choose the one with the highest probability. Finally, to label the connection, we test two methods: use the dominant application in the cluster (*Dominant* heuristic in [13], or label the connection according to the composition of the cluster and the server port it is using (*Cluster+Port* heuristic in [13]). We evaluate the efficiency of this classifier in Section 6.

5.2 Detection of the First Data Packet

After the classifier establishes that the connection is SSL, it analyzes the packets in the connection to find the first application packets. For SSLv2, the handshake can take four or six packets, depending on whether the client and the server share an encryption key that is still valid. To decide which handshake is used in a given connection, we check if the second packet sent by the client starts a key negotiation (as in figure 1).

For SSLv3, this detection is more difficult because the last packet in the SSL handshake may contain an SSL negotiation record as well as records with encrypted application payload. Therefore, to detect the first application packet in SSLv3 connections we inspect SSL records until we find the first record with content type equal to 23, which indicates an application payload. This inspection is not computationally intensive because the header of each record contains the size of the record. Figure 3 shows the distribution of the position of the first SSL packet that contains application data across all SSL connections. This result shows that there is never application data in the first two packets (which is expected from the RFC). This implies that it is safe to start inspecting TCP payloads after the third packet to identify the first application packet. This is convenient because that is what we need to detect SSL connections (as described in 5.1).

Figure 3 also shows that application data may start at any packet between 3 and 12, which justifies the need for the online packet inspection. The number of packets in the handshake depends on whether the client and the server already share a session key. Besides, a packet can consist of several SSL handshake records and SSL implementations use different methods to regroup these records.

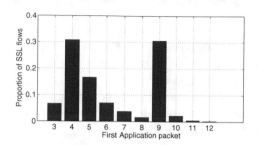

Fig. 3. Position of the first packet with application data

5.3   Application Identification

Once we have identified an SSL connection and the packets corresponding to the application, we recognize the encrypted application based on the sizes of its packets. SSL can use different encryption algorithms, which will modify packet sizes differently. Figure 4(a) shows the most common encryption algorithm used in the P6-2004 and P6-2006 traces. The specifications for SSL allows for more than 50 encryption methods. However, we can see that the five most common algorithms account for more than 98% of SSL connections.

Cipher	P6-2004	P6-2006
RC4_128_MD5 (0x04)	79.7%	66.0%
DHE-RSA-AES256-SHA (0x39)	5.9%	13.6%
AES_256_SHA (0x35)	1.0%	10.4%
RC4_128_SHA (0x05)	9.7%	7.0%
RC4_40_MD5 (0x03)	2.0%	2.1%
Other	< 2%	< 1%

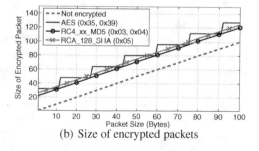

(a) Proportion of each cipher   (b) Size of encrypted packets

Fig. 4. SSL ciphers

To evaluate the influence of encryption algorithms on the sizes of exchanged packets, we design a small application that sends packets with different sizes over an SSL connection. Figure 4(b) shows the relationship between the size of application payload and the final size of the ciphered packet depending on the encryption algorithm (we focus on packets with less than 100 bytes, but the evolution remains the same for larger values). This figure shows that encryption mechanisms increase the size of the packets by an amount that depends on the cipher. For RC4 based ciphers, this increase is fixed:

25 bytes for RC4_128_SHA and 21 bytes for the other two. For AES-based ciphers, the increase varies by steps, because they use blocks.

These results are encouraging because even if encryption alters packet sizes, this change is limited and predictable. The most accurate method to decide on the size of the original packet, is to look up the encryption method in the handshake packets and transform the size of application packets accordingly. However, for the five most common ciphers this method is overkill because the increase varies from 21 to 33 bytes. Therefore, instead of keeping track of the cipher, we use a simple heuristic to decide on the size of the original packet: subtract 21 from the size of the encrypted packet regardless of the cipher.

We apply the same method of section 5.1 to the transformed sizes of the first packets with encrypted data to decide on the encrypted application. We extend the *Cluster+Port* labeling heuristic to take into account SSL-specific ports: we use 443 for HTTPS, 993 for IMAPS and 995 for POP3S.

6   Evaluation

In this section, we first evaluate our method to recognize SSL on the P6-2006 trace. Then, we validate our method to recognize encrypted applications on real Https and Pop3s traffic extracted from the P6-2006 trace and on manually encrypted connections. The test sets do not include any of the training samples.

6.1   Recognition of SSL Traffic

To evaluate the accuracy of our classifier to recognize SSL connections, we use two metrics: the proportion of connections accurately classified for all applications in our test data set (True Positives) and the proportion of connections that are wrongly labeled with each application (False Positives). Our test set consists of 50,000 connections, with more than 2,000 connections for each application. Table 2 presents both metrics for our two labeling heuristics: Dominant and Cluster+Port (defined in Section 5.1). Most misclassification errors for the Dominant heuristic happen when a connection is assigned to a cluster consisting of two (or more) applications and does not belong to the application that predominates. For all protocols the use of the Cluster+Port improves the quality of the classification because the port number helps determinate the application. Even with Cluster+Port there are some misclassifcations. These misclassifications happen when a connection in the test set follows a behavior that was not present in the training set or when a connection is assigned to a cluster with more than one application and when the server port does not help.

This table shows that our classifier, based on the sizes of the first three data packets achieves a very high accuracy and that it recognizes SSLv2 and SSLv3 for more than 80% of the connections with Cluster+Port. With the Dominant heuristic, some SSL connections are assigned to clusters which contain SSL and another application that predominates. The reason for the 2.3% of false positives for SSLv2 is that, some SSLv3 connections are classified as SSLv2 (hence the only 81% true positives for SSLv3). This is not unexpected because behaviors of SSLv2 and SSLv3 are similar in some cases.

Table 2. Application detection, including SSL(P6-2006 Trace)

Heuristice	Dominant		Cluster+Port	
Application	True Positives	False Positives	True Positives	False Positives
bittorent	74.65%	0.01%	97.30%	0.23%
edonkey	94.76%	2.89%	95.08%	0.18%
ftp	91.00%	0.04%	97.95%	0.04%
http	96.50%	2.96%	98.95%	0.00%
msn	95.36%	0.90%	100.00%	0.00%
nntp	94.40%	0.34%	99.15%	0.00%
pop3	96.65%	2.67%	99.25%	0.00%
smtp	86.35%	0.42%	98.85%	0.00%
ssh	97.73%	0.00%	96.10%	0.00%
sslv2	82.07%	2.20%	94.71%	2.30%
sslv3	67.75%	0.33%	81.20%	0.27%

However, we can easily limit the impact of this misclassification by inspecting packets from connections classified as SSLv2 to decide on the real SSL version that is used.

6.2 Recognition of Encrypted Applications

We evaluate the recognition of encrypted applications on two different test sets. First, we extract real HTTPS and POP3S connections from the P6-2006 trace: we filter SSL traffic directed to known web and mail servers from the university and obtain more than 5000 connections for both applications. Then, to evaluate our method against other applications, we manually encrypt 500 connections of FTP, Bittorent and Edonkey traffic and apply our classifier on resulting connections. We perform this classification using the same model we used in section 6.1 and therefore each connection was given a label among the 11 applications in our model.

Table 3 presents the proportion of connections correctly labeled. This table shows that applications that often use SSL (HTTP and POP3) are very well recognized when they are encrypted. The last three rows of this table evaluate our mechanism for applications that cannot be detected with port-based methods and are usually recognized based on signatures. Our classifier accurately classifies these applications when they are encrypted with more than 85% accuracy for the Cluster+port heuristic. We tested this classification without the modification of the payload size (i.e. without subtracting 21

Table 3. Detection of Encrypted Applications (HTTP and POP from P6-2006 trace, FTP, Bittorent, Edonkey manually encrypted)

Real Applications	Dominant	Cluster+Port	Manually Encrypted	Dominant	Cluster+Port
http	99.95%	99.95%	ftp	90.58%	92.67%
pop3	98.45%	98.45%	bittorent	77.87%	86.48%
			edonkey	94.56%	96.57%

bytes to each encrypted payload). The results where not as good but we were still able to achieve more than 80% accuracy for the Cluster+Port heuristic.

7   Conclusion

The contributions of this paper are two-fold. First, a **characterization of SSL usage on two campus networks**. Our analysis shows that the usage of SSL is growing and that the number of applications using SSL is increasing. Second, a **mechanism to recognize the underlying application in SSL encrypted connections based on the size of the first packets in the connections**. We show that our method achieves more than 85% accuracy. The implementation and the data used in this study are available at: http://rp.lip6.fr/ bernaill/earlyclassif.html.

In future work, we plan to extend our method to other encryption mechanisms such as SSH and IPsec. For both these protocols, the isolation of connections and the determination of the first application packets will be more challenging. Besides, the latest SSL implementations include options for compressing data and sending empty segments. These options would affect our detection mechanism and we plan to extend it to take them into account.

Acknowledgements

This study was achieved with financial support from the RNRT through the project OSCAR and from the ACI "Sécurité Informatique" through the project METROSEC.

References

1. Karagiannis, T., Broido, A., Brownlee, N., Claffy, K., Faloutsos, M.: Is p2p dying or just hiding? In: Globecom. (2004)
2. Paxson, V.: Bro: a system for detecting network intruders in real-time. Computer Networks (Amsterdam, Netherlands: 1999) **31** (1999) 2435–2463
3. Snort: http://www.snort.org.
4. Ma, Levchenko, Kreibich, Savage, Voelker: Unexpected means of protocol inference. In: Internet Measurement Confererence. (2006)
5. Song, D.X., Wagner, D., Tian, X.: Timing analysis of keystrokes and timing attacks on ssh. In: Proc. 10th USENIX Security Symposium. (2001)
6. Hintz, A.: Fingerprinting websites using traffic analysis (2002)
7. Roughan, M., Sen, S., Spatscheck, O., Duffield, N.: A statistical signature-based approach to ip traffic classification. In: IMC. (2004)
8. McGregor, A., Hall, M., Lorier, P., Brunskill, J.: Flow clustering using machine learning techniques. In: Passive and Active Measurement. (2004)
9. Zuev, D., Moore, A.: Traffic classification using a statistical approach. In: Passive and Active Measurement. (2005)
10. Moore, A., Zuev, D.: Internet traffic classification using bayesian analysis. In: Sigmetrics. (2005)
11. Erman, J., Arlitt, M., Mahanti, A.: Traffic classification using clustering algorithms. In: MineNet '06: Proceedings of the 2006 SIGCOMM workshop on Mining network data, New York, NY, USA, ACM Press (2006) 281–286

12. Bernaille, L., Teixeira, R., Akodkenou, I., Soule, A., Salamatian, K.: Traffic classification on the fly. SIGCOMM Comput. Commun. Rev. **36** (2006) 23–26

13. Bernaille, L., Teixeira, R., Salamatian, K.: Early application identification. In: To appear in Conference on Future Networking Technologies. (2006)

14. Wright, Monrose, Masson: On inferring application protocol behaviors in encrypted network traffic. the Journal of Machine Learning Research Special Topic on Machine Learning for Computer Security (2006)

15. Karagiannis, T., Papagiannaki, D., Faloutsos, M.: Blinc: Multilevel traffic classification in the dark. In: SIGCOMM. (2005)

16. Wright, Monrose, Masson: Using visual motifs to classify encrypted traffic. Workshop on Visualization for Computer Security (2006)

17. SSLv2: http://wp.netscape.com/eng/security/SSL_2.html.

18. SSLv3.0: http://wp.netscape.com/eng/ssl3/draft302.txt.

19. TLS: http://www.ietf.org/rfc/rfc2246.txt.

20. Netcraft: http://www.netcraft.com.

Measuring the Congestion Responsiveness of Internet Traffic*

Ravi S. Prasad and Constantine Dovrolis

College of Computing, Georgia Institute of Technology
{ravi,dovrolis}@cc.gatech.edu

Abstract. A TCP flow is congestion responsive because it reduces its send window upon the appearance of congestion. An aggregate of non-persistent TCP flows, however, may not be congestion responsive, depending on whether the flow (or session) arrival process reacts to congestion or not. In this paper, we describe a methodology for the passive estimation of traffic congestion responsiveness. The methodology aims to classify every TCP session as either "open-loop" or "closed-loop". In the closed-loop model, the arrival of a session depends on the completion of the previous session from the same user. When the network is congested, the arrival of a new session from that user is delayed. On the other hand, in the open-loop model, TCP sessions arrive independently of previous sessions from the same user. The aggregate traffic that the open-loop model generates is not congestion responsive, despite the fact that each individual flow in the aggregate is congestion responsive. Our measurements at a dozen of access and core links show that more than 60-80% of the traffic that we could analyze (mostly HTTP traffic) follows the closed-loop model.

1 Introduction

The stable and efficient operation of the Internet is based on the premise that traffic sources reduce their rate upon congestion. If a link has capacity C, then the offered load at that link should not exceed C for any significant period of time (relative to the maximum possible buffering in the bottleneck link). Individual TCP flows are congestion responsive thanks to the well-known TCP congestion control/avoidance algorithms. Can we expect the same, however, for streams of non-persistent TCP flows? This depends on the characteristics of the random process that generates new flows (or sessions of flows). In this paper, we rely on two well-known models to describe the flow/session arrival process at a link $\mathcal{L}$: the "open-loop" and "closed-loop" models. The former does not generate congestion responsive traffic, while the latter does. We use these models to passively estimate the congestion responsiveness of the aggregate traffic at an Internet link.

In the flow-based *open-loop model*, new flows arrive independent of the load at $\mathcal{L}$, for example according to a Poisson process. The average offered load in

---

* This work was supported by NSF CAREER award ANIR-0347374.

the open-loop model is given by λS, where λ is the average flow arrival rate and S is the average flow size. The normalized offered load is $\rho_o = \lambda S/C$. If $\rho_o < 1$ the link is stable and ρ_o is its average utilization. Otherwise, if $\rho_o \geq 1$, the link becomes unstable since the number of active flows can grow without bound. If users are impatient and abort ongoing flows after a certain time period, then the number of active flows in the underlying system will remain finite, thereby making the system stable. However, aborted flows result in user dissatisfaction and poor performance [1]. Therefore, an open loop traffic aggregate can cause instability and/or aborted flows, even if all flows use TCP. In other words, *TCP congestion control cannot avoid persistent overload when flows arrive according to the open-loop model.*

In the flow-based *closed-loop model*, we have a fixed number of users N. Each user goes through a cycle of activity, with a flow of average size S followed by an idle period of average length T_i. The average flow arrival rate is given by $\lambda_c = N/(T_t + T_i)$, where T_t is the average flow completion time. The latter depends on the load at the link, as well as on TCP (e.g., on the slow start algorithm). When $\rho_c = \lambda_c S/C \ll 1$, users spend most time thinking (i.e., $T_t \ll T_i$), and the system behaves similar to the open-loop model with arrival rate $\lambda_o = N/T_i$. However, when ρ_c approaches or exceeds 1, the number of active flows in the server increases, reducing the average per-flow throughput and increasing T_t. The increase in T_t reduces the flow arrival rate, keeping the offered load close to the capacity, i.e., $\lambda_c S \approx C$. This means that *the closed-loop traffic model is always stable and it cannot cause overload.*

A direct way to measure the congestion responsiveness of the traffic at an Internet link would be to cause a short-term congestion event, and then examine whether the flow arrival process is affected. Creating congestion events to measure the responsiveness of the traffic, however, is a highly intrusive experiment and it is often not even possible. In Section 3, we describe an indirect procedure to estimate the congestion responsiveness of the aggregate traffic at a link by classifying each session as either open-loop or closed-loop, and then measuring the fraction of traffic that follows the latter. We refer to this fraction as the *Closed-loop Traffic Ratio (CTR)*. A higher CTR value implies more congestion responsive traffic.

Our CTR estimation technique is based on the analysis of the interarrivals of packets, flows, and sessions. At this point, the technique is mostly applicable to well understood client-server applications, such as HTTP/HTTPS, FTP, news and email. An extension to peer-to-peer applications is work-in-progress. Therefore, our CTR estimates at this point cover only 30%-80% of the traffic, depending on the measured link. Measurements at a dozen of Internet links show that the CTR is usually between 60-80%. Such high CTR values suggest that a strong reason behind the congestion responsiveness of Internet traffic is that users and applications respond to congestion by slowing down the generation of new flows/sessions. TCP's congestion control and capacity overprovisioning are not the only reasons we do not see significant congestion events in most of the Internet today.

2   Congestion Responsiveness

Traditionally, congestion control has been viewed as a function of the network or transport layer at the OSI stack. The *TCP feedback loop* regulates the offered load (send-window) of a connection, based on the presence of congestion in the network (see Figure 1). The previous view, however, ignores the fact that TCP connections are the result of user and application actions. For example, the TCP connections generated from downloading a Web page, which constitute a "Web session", are the result of a user entering a URL at a web browser or clicking on a link[1]. Such connection arrivals can be consistent with an open-loop model, i.e., users generate new sessions, independent of what happened to their previous sessions and whether the network is congested or not. In other words, even though the transport layer provides congestion responsiveness through TCP, the session layer can be completely unresponsive if it keeps generating new sessions even when the network is congested.

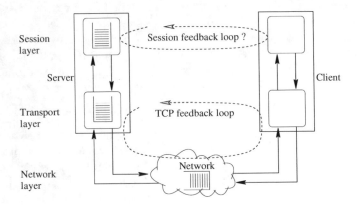

Fig. 1. The TCP feedback loop at the transport layer cannot avoid persistent overload if there is no session layer congestion control

We do not claim that TCP congestion control is not necessary. It is not sufficient, however, to avoid persistent overload. To understand this point, consider the previous example of an open-loop session layer. When the network becomes congested, each active TCP connection backs-off either reducing its send-window by a large factor or getting into a relatively long silence period (retransmission timeout). This means that congestion control pushes the offered load from each connection back to the TCP buffer of the sender. That connection is still active, however, and so it will keep trying to retransmit any lost packets and to increase its window. As the session layer keeps generating new transfers, the number of competing flows will increase, leading to a diminishing per-session goodput.

---

[1] A single click from a user can create more than one connection to download the embedded objects in a web-page. These connections together constitute one *session* and they can have different source but the same destination IP address.

TCP cannot avoid the emerging persistent overload. Instead, we need a way to tell the session (or application) layer to slow down or stop generating new flows for a while, thus forming a closed-loop system where the arrival of new flows is delayed upon congestion. Presently, this session-layer feedback is sometimes built in the application, or it simply results from the way people react when their applications are slow.

3   CTR Measurements

In this section, we propose a methodology to estimate the Closed-loop Traffic Ratio. The outline of the CTR estimation methodology is as follows. First, we partition the packet trace into a set of sessions initiated by each user. This process is far from simple, and it requires some knowledge of the corresponding application. For example, in the case of HTTP downloads, a "user" is associated with a specific IP destination address. Each session corresponds to a download operation requested by a user and it can consist of multiple "transfers". After we have transformed the packet trace into a "session trace", we then classify each session as open-loop or closed-loop. We assume that the dependency between session arrivals from the same user is related to the arrival time of the new session with respect to the finish time of the previous session from that user. Specifically, a session from a user is considered dependent on her previous session, and is classified as closed-loop, if it starts soon after the completion of the previous session. If the next session from a user starts while the previous session is still active, or after a long time from the completion of the previous session, then we view that new session as independent of the previous session and classify it as open-loop.

3.1   Definitions and Methodology

We start with a more detailed explanation of the key terms and of the CTR estimation methodology for the case of HTTP/HTTPS downloads. HTTP is not the only protocol/application for which we perform the following analysis. Traffic with other well-known ports is also well understood, in terms of who is the "user", what constitutes a "session", etc, and so we also apply the same methodology for those applications. Unfortunately, a large part of Internet traffic today does not use well-known ports. That traffic is probably generated by peer-to-peer applications. Eventually, we were able to analyze more than 50% of the TCP traffic in half of the traces we analyzed. In some traces the fraction of traffic we could analyze was as high as 78%, but in others was as low as 26%.

Users and sessions: *User* is an entity (typically a person, but it can also be an automated process) that issues Web requests. Each such application-layer request is a *session*. We assume that a user is identified in the packet trace by a destination IP address. An important exception to this rule is when multiple users share the same host (e.g., remote login) or when the addresses of different users are translated somewhere in the network to the same address (e.g., NATs or proxies). We have devised a heuristic that can identify and ignore multi-user hosts, described in the Appendix.

In HTTP/HTTPS, a user is associated with a destination address, because users typically download traffic. In applications that mostly upload traffic to remote hosts, the user is associated with a source address. A download session, associated with a certain user, can contact a number of different servers, and it can consist of several TCP connections with different destination ports. Consequently, the traffic that belongs to a certain session would have the same IP destination address, but potentially different source addresses and/or destination ports.

Connections and transfers: A TCP connection is identified in the packet trace by a unique 5-tuple field (Source IP address, Destination IP address, Source Port, Destination Port and Protocol). A connection has a start and a finish time, corresponding to the timestamps of the first and last packets in the connection, respectively. We ignore connections that were ongoing at the start or end of the trace. Pure ACKs are packets without payload. A connection with more pure ACKs than data segments is considered an *ACK flow* and it is ignored from the analysis.

HTTP 1.1 introduced persistent connections, meaning that a connection can stay alive for a long time, transferring Web objects that belong to different sessions. We partition a connection into one or more *transfers*, with different transfers being part of different sessions. Figure 2 shows an example of a persistent connection that includes two transfers. The packet interarrivals within the same

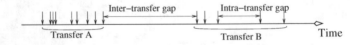

Fig. 2. Packet interarrivals from two transfers within the same connection

transfer are determined by TCP (e.g., self-clocking, retransmission timeouts) or network delays. The packet interarrivals between different transfers, however, are typically determined by the latency of user actions (e.g., clicking at a Web link). Consequently, the inter-transfer interarrivals ("gaps") are usually much longer than the intra-transfer interarrivals. We use this observation to partition a connection into transfers. If a packet interarrival within the same connection is larger than a certain *Silence Threshold* (STH), which represents the maximum intra-transfer gap, then a new transfer starts with that packet. To choose a reasonable value for STH, we examined values in the range 1sec-1min. A very small STH would partition a transfer in smaller chunks, while a very large STH would merge different transfers together. So, we expect that the average transfer size increases with STH. More importantly, we expect that for a certain range of this threshold, when it falls between the larger intra-transfer gaps and the lower inter-transfer gaps, the number of transfers is almost constant. The distribution of transfer sizes as a function of STH for a Georgia Tech inbound packet trace shows that the median transfer size is roughly constant when the threshold is between 35-45sec. In the following, we set STH to 40sec.

Grouping transfers into sessions: Since a session can consist of several transfers, as shown in Figure 3, we need to identify the transfers (TCP connections or segments of TCP connections) that were generated as a result of the same user action. The key observation here is that the interarrival of two transfers that belong to the same session will typically be much lower than the interarrival of transfers that belong to two different sessions. The latter are separated by the latency of a user action. We expect transfers of different sessions to start at least one second or so from each other, while transfers that belong to the same session are typically generated automatically by the Web browser within tens or hundreds of milliseconds.

Specifically, if the interarrival between two successive transfers is larger than a certain parameter, referred to as *Minimum Session Interarrival* (MSI), then the second transfer starts a new session. We examine the robustness of the CTR estimate to the exact choice of the MSI in Section 3.2.

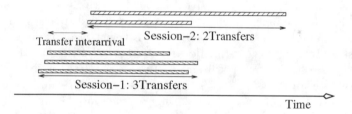

Fig. 3. Timeline of two sessions. Each session consists of several transfers.

Classifying sessions as open-loop or closed-loop: After having transformed the packet trace into a "session trace", we classify each session as open-loop or closed-loop. Recall that the key difference between these two is that in the open-loop model sessions arrive independent of the progress of previous sessions from the same user.

The first session from a user is considered open-loop, given that that session has no dependencies to previous sessions. If that user generates a new session after her previous session finishes and no later than the *Maximum Think Time* (MTT), then the arrival of this new session is considered dependent on the progress of the previous session. So, we classify that session as closed-loop. If, however, the new session arrived during her previous session or much later, more than MTT, we assume that the user either does not wait for her previous session's completion, or she was inactive for some time and now returns to the network without any "memory" of previous sessions. Thus, we classify that session as open-loop. These different cases are shown in Figure 4. We examine the robustness of the CTR estimate to the exact choice of the MTT in Section 3.2.

Other traffic with well-known ports: For other traffic, not generated by HTTP/HTTPS, we followed the convention that the transfer is an upload if the

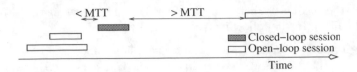

Fig. 4. Classification of different sessions from the same user as open-loop or closed-loop

destination port is well-known, and a download if the source port is well-known. The rest of the estimation methodology is the same as in Web traffic.

CTR calculation: Once we have classified sessions as open-loop or closed-loop, we then calculate the CTR as the fraction of bytes from closed-loop sessions. If the CTR is close to zero we expect that most traffic is congestion unresponsive (open-loop model), while if the CTR is close to one we expect that most traffic would reduce its flow arrival rate upon congestion (closed-loop model).

Limitations: The main assumption in the previous methodology is that the timing between the arrivals of successive sessions from the same user can reveal whether sessions arrive independently of each other. This is not always the case of course. For example, a user can obtain a URL through an ongoing web-page download (session X) and then start downloading that URL in another window (session Y), while session X is still active. In this case, the arrival of session Y depends on the progress of session X. However, session Y starts before X completes, and so we would classify the former as open-loop. On the other hand, in a fast and uncongested network, even though a user can start a sequence of independent sessions, some of the sessions can complete before a subsequent session starts. In this case, the latter will be incorrectly classified as a closed-loop session. We expect that such "deviations" are not common compared to the more common Web browsing behavior, which is to first download a page (open-loop session arrival), spend some time reading or viewing it, and then either download another page (closed-loop session arrival) or leave the system for a while.

3.2   Robustness of Estimation Methodology

The CTR estimate depends on the following parameters: STH, MSI, and MTT. We have already mentioned that a robust value for STH is around 40sec. In this section, we investigate the optimal range for MSI and MTT, and examine the CTR's robustness to the choice of these two parameters.

The MSI is used to merge together different transfers of the same session. As these transfers are typically machine-generated, they start almost simultaneously. In the packet trace, however, they can appear with slightly longer spacings due to network delays. A reasonable range for this parameter is between 0.5-1sec. The MTT, on the other hand, represents the longest "thinking" time for a user during Web browsing, before we assume that that user left the system. A reasonable range for this parameter is between 5-30min.

We examined the CTR variations for a Georgia Tech trace with different MSI and MTT values. The main finding is that the CTR does not significantly depend on these parameters (it varies in a small range between 0.6-0.72) as long as the MSI and the MTT fall between 0.5-2sec and 5-25min, respectively. To further examine the robustness of CTR against these two thresholds, we also performed two-factor analysis of variance. The null hypothesis is that the CTR is independent of these two parameters. We can reject this hypothesis at a significance level of 0.05. However, the hypothesis that the CTR is independent of the interaction of these two parameters cannot be rejected at the same significance level. We found that the slope of the CTR with respect to MSI is 0.0232/sec, and with respect to MTT it is 0.0020/min. Therefore, we concluded that the CTR is practically insensitive to these parameters in the ranges that we consider. In the rest of the analysis, we set MSI=1sec and MTT=15min.

Table 1. CTR of the analyzed TCP traffic at various links

Trace_ID	Direction	Collection time	Link location	Duration	TCP traffic Total GB (%)	Well-known ports bytes	CTR
GaTech-in	In	07-Jan-05	GaTech	2Hr.	129.74 (97.15)	63.50%	0.71
GaTech-out	Out	07-Jan-05	GaTech	2Hr.	208.06 (98.92)	47.90%	0.57
Los-Nettos	Core	03-Feb-04	Los-Nettos, CA	1Hr.	59.37 (94.96)	65.59%	0.77
UNC_em0	Out	29-Apr-03	UNC	1Hr.	153.19 (97.33)	35.78%	0.61
UNC_em1	In	29-Apr-03	UNC	1Hr.	41.51 (87.76)	26.59%	0.76
UNC_em1_2	In	24-Apr-03	UNC	1Hr.	55.25 (85.67)	44.88%	0.78
MFN_0	Core	14-Aug-02	MFN, San Jose	1Hr.	151.38 (96.31)	61.09%	0.69
MFN_1	Core	14-Aug-02	MFN, San Jose	1Hr.	186.93 (97.75)	71.83%	0.62
IPLS_0	Core	14-Aug-02	Abilene	1Hr.	172.22 (96.40)	41.93%	0.70
IPLS_1	Core	14-Aug-02	Abilene	1Hr.	177.99 (85.00)	47.27%	0.64
Auckland_0	In	06-Nov-01	Auckland, NZ	6Hr.	0.58 (94.83)	72.99%	0.73
Auckland_1	Out	06-Nov-01	Auckland, NZ	6Hr.	1.44 (98.43)	77.99%	0.67

3.3 Results

Table 1 summarizes the metadata of the traces we analyzed in this paper. The traces are from university access links, commercial access links and backbone links, and they were collected between 2001-2005. It also shows the CTR estimates for Web and other well-known port traffic in these 12 Internet traces.

An important observation is that the CTR for access links is always higher in the inbound direction than in the outbound direction. This can be explained by the fact that users that initiate sessions in the inbound direction belong to the limited population of users within that campus network. On the other hand, users that initiate sessions in the outbound direction come from all over the Internet and they belong to a much larger population. Consequently, the fraction of open-loop traffic in the latter is higher (lower CTR).

The most important observation, however, is that the CTR for almost all traces is high, typically between 60-80%. Even the backbone links, where we would expect more open-loop traffic due to the large number of users, have a high CTR. This suggests that a major reason for the congestion responsiveness of Internet traffic may be that most applications follow the closed-loop model, and so they are responsive to congestion at the session generation layer.

4   Related Work

The notion of congestion responsiveness is related to the issue of TCP-friendliness [2]. The latter, however, focuses on non-TCP traffic. Previous studies with open-loop flow arrivals concluded that admission control is necessary to avoid persistent overload [3,4]. Berger and Kogan [5], as well as Bonald et al. [6], used a closed-loop model to design capacity provisioning rules for meeting throughput-related QoS objectives. Heyman et al. [7] used a closed-loop model to analyze the performance of Web-like traffic over TCP, showing that the session goodput and fraction of time the system has a given number of active sessions are insensitive to the distribution of session sizes and think times. In a recent paper, Schroeder et al. [8] compare open-loop and closed-loop arrival models and highlight their differences in terms of mean job completion time and response to different scheduling policies.

Understanding user and application behavior, as well as their impact on network traffic, has been the subject of some previous work [9,10,11,12]. These papers focus on HTTP, mostly because the application characteristics of Web browsing are well understood. Casilari et al. [9] present models for different levels of HTTP analysis, namely at the packet, connection, page or session levels. Smith et al. attempt to reconstruct HTTP sessions from unidirectional TCP header traces [11]. Feldmann [10] correlates bidirectional IP traces with HTTP header traces. Tran et al. [12] use payload information to identify related HTTP requests/responses and to study the effect of congestion on user behavior (probability of aborted and retried sessions).

5   Discussion and Future Work

This paper focused on a simple question: how does Internet traffic react to congestion? This is an important question towards a better understanding of the Internet traffic characteristics. Additionally, the issue of congestion responsiveness has several practical implications. To avoid overload conditions, applications need to be "congestion-aware" at the session layer. One way to do so is to delay generating a new session to a server if the previous has not yet completed or if it is stalling. If the traffic at a certain link is mostly open-loop, then the operator has two options. One is to increase the link capacity above the given offered load. The second is to do some sort of admission control at that link. With closed-loop traffic, on the other hand, these two options may not be necessary because the offered load is self-regulating. We are currently extending this work to estimate the CTR for peer-to-peer traffic, which often accounts for most of the traffic today.

References

1. Yang, S.C., de Veciana, G.: Bandwidth Sharing: The Role of User Impatience. In: Proceedings IEEE GLOBECOM. (2001) 2258–2262
2. Floyd, S., Handley, M., Padhye, J., Widmer, J.: Equation-Based Congestion Control for Unicast Applications. In: SIGCOMM. (2000)
3. de Veciana, G., Lee, T., Konstantopoulos, T.: Stabily and Performance Analysis of Networks Supporting Services. IEEE/ACM Trans. on Networking **9** (2001)
4. Fredj, S.B., Bonald, T., Proutiere, A., Regnie, G., Roberts, J.W.: Statistical Bandwidth Sharing: A Study of Congestion at Flow Level. In: Proceedings of ACM SIGCOMM. (2001)
5. Berger, A., Kogan, Y.: Dimensioning Bandwidth for Elastic Traffic in High-Speed Data Networks. IEEE/ACM Transactions on Networking **8** (2000) 643–654
6. Bonald, T., Olivier, P., Roberts, J.: Dimensioning High Speed IP Access Networks. In: 18th International Teletraffic Congress. (2003)
7. Heyman, D., T.V.Lakshman, Neidhardt, A.L.: A New Method for Analysis Feedback-Based Protocols with Applications to Engineering Web Traffic over the Internet. In: ACM SIGMETRICS. (1997)
8. Schroeder, B., Wierman, A., Harchol-Balter, M.: Closed Versus Open System Models and their Impact on Performance and Scheduling. In: Symposium on Networked Systems Design and Implementation (NSDI). (2006)
9. Casilari, E., Reyes-Lecuona, A., Diaz-Estella, A., Sandoval, F.: Characterization of Web Traffic. In: IEEE Globecom. (2001)
10. Feldmann, A.: BLT: Bi-Layer Tracing of HTTP and TCP/IP. Computer networks **33** (2000) 321–335
11. Smith, F., Campos, F., Jeffay, K., Ott, D.: What TCP/IP protocol headers can tell us about the Web. In: ACM SIGMETRICS. (2001)
12. Tran, D.N., Ooi, W.T., Tay, Y.C.: SAX: A Tool for Studying Congestion-Induced Surfer Behavior. In: PAM. (2006)

Appendix: Multi-user Host Detection

We describe a heuristic to distinguish between single-user and multi-user hosts (such as NATs, proxies, rlogin servers etc). A host is identified by a unique IP address in the packet trace. In a multi-user host, sessions generated by different users share the same destination address, making it impossible to distinguish sessions from different users. In multi-user hosts, however, the number of transfers per session would typically be much larger than in single-user hosts. This large difference is the key criterion to detect multi-user hosts. For instance, in one of the Georgia Tech inbound packet traces about 99% of hosts have less than 20 transfers per session. However, there are also 100 hosts (<1%) that generate up to a few thousands of transfers per session; it is likely that they are multi-user hosts. We examined the DNS names of those hosts, and several of them indicate proxies and firewalls. So we chose a threshold of 10 transfers per session, on the average, to distinguish single-user from multi-user hosts. It turns out that the final CTR estimate (which is what we care about) is robust to the previous threshold, as long as the latter is more than 5-6 and less than about 100 sessions.

Profiling the End Host

Thomas Karagiannis[1], Konstantina Papagiannaki[2],
Nina Taft[2], and Michalis Faloutsos[3]

[1] Microsoft Research
[2] Intel Research
[3] UC Riverside

Abstract. Profiling is emerging as a useful tool for a variety of diagnosis and security applications. Existing profiles are often narrowly focused in terms of the data they capture or the application they target. In this paper, we seek to design general end-host profiles capable of capturing and representing a broad range of user activity and behavior. We first present a novel methodology to profiling that uses a graph-based structure to represent and distill flow level information at the transport layer. Second, we develop mechanisms to: (a) summarize the information, and (b) adaptively evolve it over time. We conduct an initial study of our profiles on real user data, and observe that our method generates a compact, robust and intuitive description of user behavior.

1   Introduction

Profiling a behavior refers to the act of observing measured data and extracting information which is representative of the behavior or usage patterns. Profiling is useful in developing a model of the behavior and in deriving guidelines of what is normal and abnormal within that context. Examples of successful uses of profiles include profiling of traffic patterns on server links to uncover DoS and flash crowd events [4], web-server profiling [11], power usage profiles for efficient power management [10], profiling end-to-end paths to detect performance problems [8], profiling of traffic patterns on aggregated gateway and router links to facilitate accurate application classification [5], etc.

While there has been research on profiling web server traffic [4,11], and gateway and backbone links (i.e., highly aggregated traffic) [5,12], end-host profiling has received little attention. One work in this area is [7] in which the authors build end-host profiles with the goal of defending against worm attacks. Their profile describes the community of hosts an end-system normally interacts with.

We believe that observing the host behavior at the transport layer can reveal a wealth of information, such as: behaviors on who tries to talk to the host, who the host communicates with, the mix of applications used, the dispersion (or randomness) of the destinations contacted for a particular application, the pattern of port usage, the evolving mix of protocol usage, and so on. A number of security applications have identified particular features, derivable from packet header fields, as useful for detecting specific attacks. For example, many IDS

S. Uhlig, K. Papagiannaki, and O. Bonaventure (Eds.): PAM 2007, LNCS 4427, pp. 186–196, 2007.

systems will declare a machine compromised if the number of simultaneous TCP connections exceeds a predefined threshold [2,1]. Yet other systems look for a change in the dispersion on destination IP addresses to find anomalies [6].

All aforementioned work tends to define profiles within the bounds of their intended use. In this work, we are trying to formalize the concept of profiling transport layer information and identify desirable properties. We believe that a profiling mechanism, focused on end-hosts, should meet the following goals.

- *Goal 1:* It should be able to identify dominant and persistent behaviors of the end-system, capturing repeatable behaviors over time.
- *Goal 2:* It should be a compact enough representation, avoiding excess detail that may correspond to ephemeral behavior. This is important in a networking setting due to the use of ephemeral ports by certain applications.
- *Goal 3:* It should be stable over short-time scales avoiding transient variability in the host behavior.
- *Goal 4:* It should evolve by adding new behaviors and removing stale ones.
- *Goal 5:* It should be able to capture historical information to illustrate typical ranges of values for features.

Our main contribution is a novel approach to profile end-host systems based on their transport-layer behavior. First, we propose the use of a graph-based structure which we call a *graphlet*, to capture the interactions among the transport-layer protocols, the destination IP addresses and the port numbers. Note that *graphlets* have two key properties: (a) they are extensible, since through annotations of the nodes or links one could achieve a lossless representation of all flow information through a single graph, and (b) they provide intuitive and interpretable information. The notion of *graphlets* was introduced in [5] for application classification. Building on [5], we extend this original idea in a number of nontrivial ways. Second, we design a two step method that is based on an unsupervised online learning process. In the first step, we build and continuously update *activity graphlets* that capture all the current flow activity. The second step contains mechanisms to (a) compress the large *activity graphlets* to retain only essential information, and (b) to evolve this latter summary in a way that reflects changes over time. The output of this process is called a *profile graphlet*.

Using enterprise network traces, we find that user activity is successfully captured in our profiles. In particular, our profiles capture roughly 70-90% of all user activity, yet are about 80-90% smaller in size relative to the uncompressed activity graphlets. This result demonstrates that our profiles are efficient and compact while still remaining highly descriptive. Our initial findings indicate that profiles can vary greatly across users and this motivates the use of end-host profiles for security, diagnosis and classification applications. One of our interesting findings is that over short time scales (e.g., 15 minutes) the profiles evolve slowly typically experiencing small changes, yet over longer periods of time (e.g., a month), the majority of the profile content may change. This indicates that most parts of the profile are apt to change, and further underscores the need for adaptivity.

2 Data Description

We collected packet header traces within a secure enterprise network environment. Using the *CoMo* monitoring tool [9], we capture all traffic on the access link of our office building. Two traces were collected; one spans the entire month of October 2005, and the other a two week period in November 2005. We monitor the traffic of roughly 200 distinct internal IP addresses, that collectively represent user laptops and desktops, as well as network infrastructure equipment (e.g., NFS or DNS servers).

3 Methodology

3.1 Capturing Host Activity Via Graphs

A fundamental element of our profiling methodology is the special purpose graph, called **graphlet**. The concept of *graphlets* was first introduced by the *BLINC* methodology [5] to capture the distinct transport layer footprint of different applications, termed as "application *graphlets*". Application *graphlets* were further used for the classification of the traffic observed at a traffic aggregation point into applications. Our use of *graphlets* in this work is significantly different from the original BLINC work. We extend the definition of a graphlet, introduce graphlet annotations and manipulate the graphlet in different ways (in terms of learning, updating, compacting, etc). First, the intended goal and use of *graphlets* in our work is substantially different compared to BLINC. BLINC's goal was to identify application footprints in traffic streams in a *supervised manner* according to the pre-defined *application graphlets*. On the contrary, we study *graphlets* with the goal to profile hosts in an *unsupervised way*, i.e., we learn (and update) an unknown user behavior on-line. Second, as described below, we extend the graphlet definition to include additional elements. Third, we introduce ideas for summarizing and creating compact *graphlets*.

A *graphlet* is a graph arranged in six columns corresponding to: (srcIP, protocol, dstIP, srcport, dstport, dstIP). Fig. 1(top left) presents an example graphlet that was derived from one of our *LAB* hosts and plotted using the graphviz tool [3]. The BLINC *graphlets* consisted only of the first 5 columns; the additional sixth column here is critical to our methodology.

Each **graphlet** node[1] presents a *distinct* entity (e.g. port number 80) from the set of possible entities of the corresponding column. The lines connecting nodes imply that there exists at least one flow whose packets contain the specific nodes. This way, each flow creates a directed **graphlet** path starting from the host IP address on the left and traversing the appropriate entities in each column. Note that we define a **flow** by the 5-tuple of the packet header, and the flow can consist of one or more packets. Similarly, a *graphlet* path can correspond to a multitude

---

[1] The term node indicates the components of a *graphlet*, while the term host indicates a communicating device.

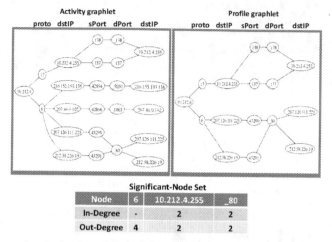

Fig. 1. Example of a host activity and profile graphlets and the significant node set

of flows with the same 5-tuple definition. The destination IP appears twice at the third and sixth column in the *graphlet*. This redundancy is critical, since it allows us to observe *all* pairwise interactions between the most information-heavy fields of the 5-tuple: destination IP address ($dstIP$), the source port ($srcport$) and destination port ($dstport$).

If many flows traverse a node, the node will most likely have a high degree. By construction, all edges in a *graphlet* are between nodes of adjacent columns. If we traverse a *graphlet* from left to right by following a path, we define a direction in visiting the nodes. This way, we can define the **in-degree (out-degree)** of a node as the number of edges on the left (right) side of the node. The in- and out-degree of a $dstIP, srcport, or dstport$ node abstracts its interaction with the other two types of nodes. For example, the out-degree of a node representing port 80, captures the dispersion of addresses visited using web applications.

Because we are building profiles for a single host, there is only one source IP address and hence this field is not included in what we are terming the "heavy information fields" of the 5-tuple. We point out that a graphlet is a directed graph. When the host is the source, the directed edges flow from left to right in our depictions. If the directed edges flow from right to left, then our host is the recipient of incoming flows. Note that although conceptually each profile consists of two directed graphs, in practice a single data structure can be designed to capture all the needed information.

For example, Fig. 1 presents an "activity" graphlet which resulted by observing all the incoming and outgoing flows of a host during a specific time window. The "profile" graphlet refers to our distilled and compact version of the activity graphlet. (Activity and profile graphlets, along with significant node sets are discussed in Sec. 3.3).

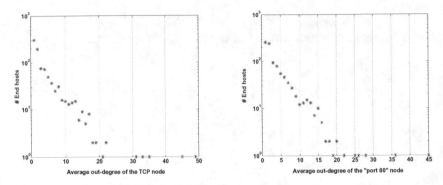

Fig. 2. Histograms of the average out-degree of two different nodes (TCP and "port 80" node) in the client *graphlets* computed every 15 minutes. Significant variations in the number of out-degrees across clients point towards client personalized profiles.

3.2 The Advantages of *graphlet* Profiling

We believe that graphlets are an interesting approach to end-host profiling for a number of reasons. First, one could imagine keeping per host flow records in order to compute statistics regarding its behavior. A database of flow records is an enormous amount of information. Instead, our graphlet achieves a representation of important information in a compact form limiting the redundancy. Second, such flow records are not interpretable without further processing. However, the paths, nodes and node properties in graphlets are easy to interpret.

We can further expand *graphlets* to annotate nodes with temporal information. For example, we can create time series information for each node (e.g., the time series of the out-degree). This is equivalent to annotating the nodes in the graph and tracking the evolution of the weights. Similarly, we can attach weights to links in the graph in order to track more typical features, such as the number of packets or bytes for all flows transiting that path. Existing security solutions use threshold based-techniques on metrics like the number of TCP connections per destination port per time interval. Recent solutions examine the dispersion of the 3 key fields [6]. All such techniques can be captured within the framework of weight-annotated *graphlets*. The power of this profiling mechanism is that it goes beyond these methods, since it also incorporates the graph relationships, all in a single structure. We illustrate this here with three examples:

- The out-degree of the TCP node (or any protocol node) reveals the typical number of TCP destination IPs per client. By observing how the out-degree of the TCP node in the graphlet evolves over time, we learn about the typical range for the number of simultaneous destinations contacted through TCP within a window of time (the time scale of the graphlet). For example, Fig. 2 (left) presents a histogram of the average *out-degree* of the TCP node for all our client graphlets every 15 minutes. We observe a wide range of behavior.
- For applications with well-known port numbers, *graphlets* can reveal what is the typical number of destination IPs contacted for each given application. For

Method: Construct Profile
1. Upon arrival of each packet, update *activity graphlet* if flow information not already included.
2. Every *t* minutes
   a). identify new significant activity, according to **summarization policy** as candidate to join profile graphlet.
   b). Add new significant activity into profile, if approved by **delayed-accept policy**, using Algorithm 1.
   c). Remove stale parts of profile according to **aging policy**.

Fig. 3. Summary of Method

Algorithm 1: Populate Profile Graphlet with Significant Nodes
Repeat until all significant nodes processed
   1. Rank all nodes in *activity graphlet* according to their maximum in-degree or out-degree:
   $max\{indegree, outdegree\}$
   2. Remove the highest degree node and all its edges. Insert into *profile graphlet*.

Fig. 4. Algorithm: inserting significant nodes into profile

example, examining the out-degree of the graphlet node for destination "port 80" reveals the number of destinations typically contacted by an HTTP application. Similarly Fig. 2 (right) presents a histogram of the average *out-degree* of the "port-80" node for all our client graphlets computed every 15 minutes. Again we see considerable variability across hosts.

- Scanning behavior can be easily seen from graphlets. For example, port scanning would appear as an excessively large number of destination ports associated with a single destination address. Similarly if a host initiates an address-space scan for a specific port (worm-like behavior) this would appear as an excessively large number of destination IPs associated with a single destination port.

3.3 Building Profiles

All the information obtained from monitoring a host's communication traffic could lead to an enormous graphlet (called the *activity graphlet*). Recall that, as per our goals described in the introduction, our aim is to capture "typical" or "persistent" behaviors in a compact way that avoids transient noise. We now describe our methods for converting the activity graphlet into a profile graphlet via policies for compression (i.e., summarization) and adaptivity.

Our method is depicted in Fig. 3. The policies used in this method were designed as follows. The intuition behind our **summarization policy** comes from observations on our trace data that activity graphlets do vary dramatically from one host to another, and a summary metric such as *number of nodes in the graphlet* is very volatile. Looking at activity *graphlets* across many hosts, time intervals and traces, we did find one common characteristic; namely that they feature a small number of high degree nodes ("knots" in the *graphlet*). These nodes result from flows that share at least one *graphlet* node (e.g., distinct web flows that share port 80). At the same time, our activity *graphlets* featured a number of paths comprising only one-degree nodes (ignoring protocol nodes).

See for example the middle two paths in Fig. 1(top left). Typically, those corresponded to ephemeral flows[2].

Building on this insight, we define the set of **significant nodes** in an activity *graphlet* to be those nodes with an *in-degree or out-degree larger than 1*. The only nodes we retain in our graphlet profiles are the significant nodes. We populate our graphlet profiles using the procedure outlined in Fig. 4. Fig. 1 gives an example of an activity graphlet and the resulting profile graphlet that it generates. We use the term **significant set** to refer to the group of significant nodes of a graphlet. The **profile** *graphlet* consists of the union of all the flows that are affiliated with the significant nodes. As such, the profile *graphlet* is a subset of the initial activity graphlet. Thus, we could say that our profiling consists of two components: (a) the significant set, and the (b) profile *graphlet*.

In order to evolve, our profiles need to: 1) remove information when it becomes stale, and 2) add new content when it becomes relevant. The time scale of this adaptivity affects both the stability and meaningfulness (i.e., utility) of the profile. If the profiles evolve too quickly, they will be less stable (nodes will be added and removed very frequently); whereas if they evolve too slowly, they will be less meaningful (miss new important nodes and contain stale ones). Let t denote the update period of the profile graphlet. Updating the profile means that the set of significant nodes at a time instance t is the union of the sets at time $t-1$ and t.

We employ a **delayed-accept policy** to control the addition of new nodes. Significant nodes are not inserted in the graphlet profile unless they are active for at least two consecutive intervals t. Such a mechanism is robust to ephemeral nodes introduced by the reuse of port numbers across flows.

We make use of an **aging policy** to remove obsolete information. A significant node is removed from a profile if it is inactive for some period of time. Our timeout period N is measured in days. Inactivity refers to nodes that are currently not significant but were in previous time intervals. Due to space constraints, we do not illustrate the stability and utility tradeoffs we observed for various values of t and N. In short, we found that using an update period t equal to 15 minutes, and aging threshold N of one week achieved a good tradeoff between utility and stability.

4   Properties of the End-System Profiles

Here we describe the properties that establish the robustness of significant nodes as a means of profiling end-user activity. To this end, we examine the extent to which our profiles meet the five goals mentioned in section 1.

Goal 1 - Capturing representative information: We first examine the identities of the nodes that populate the user profiles. Intuitively, the nodes should depict the primary activities of each end-system and if possible also reflect its functional role in the network (e.g., client vs. server).

---

[2] Note that ephemeral flows refer to a whole path in the *graphlet*, while ephemeral nodes only to the specific node.

Table 1. Profile instances of various end-systems

Host	activity *graphlet* size	significant node set in the profile
Client1	104	dst ports: 22 (SSH), 443 (HTTPS), 80 (HTTP), 2233 (VPN)
Client2	72	dst ports: 993 (IMAP), 137 (NETBIOS), 80 (HTTP), 995 (POP3)
Client3	259	dst ports: 80 (HTTP), 6881, 6882, 6884, 6346, 16881 (P2P)
NFS SERVER	31	src port: 2049 (NFS)
LDAP SERVER	309	src ports: 389 (LDAP), 139 (NETBIOS)

Table 1 presents five profile instances for three randomly picked clients and two servers from our enterprise networks. We observe that *all* significant nodes in the client profiles are destination ports reflecting well-known services accessed by the clients. Note that client 3 appears to run the *BitTorrent* peer-to-peer application and the set of significant nodes reflects common ports used by this application. The significant nodes for the servers, however, reflect the ports where the offered services reside.

Table 2. Most popular significant nodes

dstP = 80	dstP = 5499	dstP = 443	dstP = 2233	dstP = 53	dstP = 1863	dstP = 389	dstP = 22
WEB	CHAT	HTTPS	VPN	DNS	MSN	LDAP	SSH

To examine the identities of our profiles in a broader setting, we looked at the most popular significant nodes across all profiles. Table 2 presents the eight most popular nodes which, similarly, represent services at well-known ports. This initial data exploration indicates that *our profiles are able to capture dominant and meaningful end-system behavior and discriminate its functional role in the network.*

Note that while a number of significant nodes are common in host profiles, these significant nodes can be annotated with a variety of information such as their average out-degree to capture the user variability as shown in Fig. 2.

Goal 2 - Compact representation: To assess the breadth and compactness of the profiles, we define two metrics. *Compression* is defined as the ratio of the number of significant nodes over the total number of nodes in the activity graphlet. *Coverage* is defined as the fraction of flows that the profile captures compared to the total number of the flows generated by the host. (A flow is defined here as a unique 5-tuple.) A good profile should achieve high coverage and high compression because the significant nodes should: a) represent the majority of the activity of the edge-host (high coverage), b) amount to only a small number of the total nodes in the *graphlet* (high compression).

Fig. 5(left) shows that abstracting the *graphlet* to a set of significant nodes leads to a compression greater than 80% compared to the activity *graphlet*. We also see that the significant nodes often cover more than 90% of the flows sourced at the host. Recall that by definition, "uncovered" flows correspond to those whose path comprises only one-degree nodes in the *graphlet*. We thus conclude that *our set of significant nodes offers both high compression and coverage.*

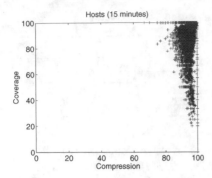
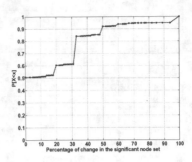

Fig. 5. LEFT: Coverage vs. compression for hourly *graphlets* (LAB trace). RIGHT: Similarity across consecutive intervals for the significant node set. For approximately 50% of the intervals the significant node set remains the same across time.

Goal 3 - Stability: Recall that our profiles are updated every 15 minutes. We now examine the amount and nature of changes occurring in the profile over time. For each end host, we examined the difference in the set of significant nodes from one time slot to the next. The difference is the ratio of the number of nodes present in both intervals divided by the average number of significant nodes in the two sets. In Fig. 5(right) we present the CDF of these ratios for all hosts over all time slots.

We observe that roughly 50% of the time there is no change at all from one time slot to the next. Also, less than 10% of the time does a node change by more than 70%. We conclude that while there appears to be a reasonable amount of stability from one 15 minute window to the next, every so often the profile can experience a large change. These initial results hint that perhaps over shorter time scales these profiles can remain stable, yet over longer time periods, profiles can experience large amounts of change. This indicates that a notion of stability should perhaps be tied to the amount of evolution occurring in user behavior. This is a subject of our future research.

Goal 4 - Evolvability: Fig. 6(left) demonstrates the impact of the "delayed-accept" and "aging" policies on the total number of significant nodes for all hosts in the network. The upper line corresponds to the total number of significant nodes across all hosts when only "aging" is used, while the bottom line also incorporates the effect of "delayed-accept". During the first week of profiling the number of significant nodes shows a constant increase in both cases. This is the "learning" stage of our approach and lasts approximately 1.5 weeks. While the effect of "delayed-accept" is evident across time, "aging" is observed after the first week due to our choice of "weekly" history. The sum of significant nodes appears not to vary significantly after approximately 2 weeks. Note that while the time interval on the *x-axis* spans a time period of a month, we only observe a few changes. These initial results indicate that our *delayed-accept* and *aging* policies do manage to filter transient behavior while balancing the stability.

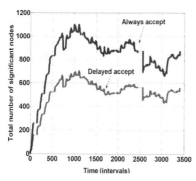
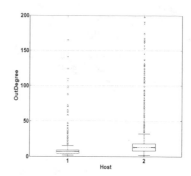

Fig. 6. LEFT: Total number of significant nodes for all network hosts when we use delayed-accept" and "aging". RIGHT: Boxplot of the outdegree time-series for a common significant node for two host profiles.

Goal 5 - Capturing historical information: Recall from section 3 that each significant node in the profile can be annotated with various time-series information. Fig. 6(right) presents such an example in a box plot showing the out-degree of a common significant node (web) across two hosts. Such time-series can be further analyzed to provide insight regarding *typical* individual behavior (e.g., average number of TCP connections), a *range* of behavior (e.g., 90 percentile points), and *outliers* (denoted with the points outside the wedges in the figure). We postulate that this sort of information could be important for anomaly detection applications (benign or malicious).

5   Conclusions–Discussion

In this paper, we present a novel approach to profile end-host systems based on their transport-layer behavior. We introduce the idea of using graphs to capture flow information and inter-flow dependencies. We illustrate that all of a host's flow data can be greatly compressed into a compact representation, that captures dominant user behavior. Initial results suggest that a user's behavior can undergo large changes over time, and this underscores the need for adaptive profiling.

We envision our profiling methodology being used in many different ways depending on the intended goal. Examples include:
• For enterprise network management, to understand user behavior for resource provisioning, load balancing, allowing for user clustering based on similar profiles, etc.
• Monitoring the profile graphlet in comparison to the activity graphlet could be useful for anomaly detection. Abrupt changes in either the normal range of behavior, or outlier events, could signal an anomaly, whether benign or malicious.
• Monitoring the patterns in the out-degrees of protocol-nodes, or other significant nodes, could reveal scanning attempts.

References

1. Intrusion Detection Systems (IDS) Part 2 - Classification; methods; techniques.   In *http://www.windowsecurity.com/articles/IDS-Part2-Classification-methods-techniques.html*, 2004.
2. Arbor Networks. http://www.arbor.net/.
3. Graphviz. http://www.graphviz.org/.
4. J. Jung, B. Krishnamurthy, and M. Rabinovich. Flash crowds and denial of service attacks: Characterization and implications for cdns and web sites. In *Proceedings of the 11th International World Wide Web Conference*, May 2002.
5. T. Karagiannis, K. Papagiannaki, and M. Faloutsos. BLINC: Multi-level Traffic Classification in the Dark. In *ACM SIGCOMM*, August 2005.
6. A. Lakhina, M. Crovella, and Christophe Diot. Mining Anomalies Using Traffic Feature Distributions. In *Proc. of ACM SIGCOMM*, August 2005.
7. P. McDaniel, S. Sen, O. Spatscheck, J. Van der Merwe, B. Aiello, and C. Kalmanek. Enterprise Security: A Community of Interest Based Approach. In *Proc. of Network and Distributed System Security (NDSS)*, Feburary 2006.
8. V. Padmanabhan, S. Ramabhadran, and J. Padhye. NetProfiler: Wide-Area Networks Using Peer Cooperation. In *Proceedings of the Fourth International Workshop on Peer-to-Peer Systems (IPTPS)*, February 2005.
9. The CoMo Project. http://como.intel-research.net/.
10. G. Theocharous, S. Mannor, N. Shah, B. Kveton, S. Siddiqi, and C.-H. Yu. Machine Learning for Adaptive Power Management, 2006. Intel Technology Journal.
11. Mengjun Xie, Keywan Tabatabai, and Haining Wang. Identifying Low-Profile Web Server's IP Fingerprint. In *IEEE QEST*, 2006.
12. K. Xu, Z.-L. Zhang, and S. Bhattacharyya. Profiling Internet Backbone Traffic: Behavior Models and Applications. In *ACM Sigcomm*, August 2005.

A Method to Estimate the Timestamp Accuracy of Measurement Hardware and Software Tools

Patrik Arlos and Markus Fiedler

Blekinge Institute of Technology, School of Engineering,
Dept. of Telecommunication Systems, Karlskrona, Sweden
{patrik.arlos,markus.fiedler}@bth.se

Abstract. Due to the complex diversity of contemporary Internet applications, computer network measurements have gained considerable interest during the recent years. Since they supply network research, development and operations with data important for network traffic modelling, performance and trend analysis etc., the quality of these measurements affect the results of these activities and thus the perception of the network and its services. One major source of error is the timestamp accuracy obtained from measurement hardware and software. On this background, we present a method that can estimate the timestamp accuracy obtained from measurement hardware and software. The method is used to evaluate the timestamp accuracy of some commonly used measurement hardware and software. Results are presented for the Agilent J6800/J6830A measurement system, the Endace DAG 3.5E card, the Packet Capture Library (PCAP) either with PF_RING or Memory Mapping, and a RAW socket using either the kernel PDU timestamp (`ioctl`) or the CPU counter (TSC) to obtain timestamps.

1 Introduction

In recent years computer network measurements have gained much interest, one reason is the growth, complexity and diversity of network based services. Computer network measurements provide network operations, development and research with information regarding network behaviour. The accuracy and reliability of this information directly affects the quality of these activities, and thus the perception of the network and its services. References [1,2] provide some examples on the effect that accuracy has on bandwidth estimations.

One major source of error is the timestamp accuracy obtained from measurement hardware and software. Therefore, in this paper we present a method that estimates the timestamp accuracy obtained from measurement hardware and software. The method has been used to evaluate the timestamp accuracy of some commonly used hardware, Agilent J6800/J6830A and Endace DAG 3.5E [3,4], and software, Packet Capture Library [5] and derivatives (PF_RING [6] and Memory Mapping [7]) as well as a raw socket using either `ioctl` or TSC [8] to obtain PDU timestamps. The software was evaluated using different hardware platforms, operating systems as well as clock synchronisation methods. We have

S. Uhlig, K. Papagiannaki, and O. Bonaventure (Eds.): PAM 2007, LNCS 4427, pp. 197–206, 2007.

intentionally used off-the-shelf components, as there are many measurements performed using such equipment and software. Furthermore, in [9] we evaluated the accuracy of DAG [3], Tcpdump [5] and Windump [10] when the tools stored the trace to disc, we found that Tcpdump reported PDUs with wrong timestamp, and that Windump lost PDUs without reporting the loss to the user, the PDU inter-arrival times of both Tcpdump and Windump showed large tails, which merrited further investigation. All in all, it was obvious that the accuracy of the tools was clearly affected by the way the data was processed. Hence, in this paper we've focused on the measurement system, not the system that stores the data to disc, we do this by using a Distributed Passive Measurement Infrastructure [11] which separates the measurement task from the analysis and visualization tasks.

The outline for this paper is as follows. In Section 2 we will discuss timestamp accuracy, this is followed by Section 3 where we describe the method. Section 4 describes the measurement setup and in Section 5 we evaluate the timestamp accuracy for a few common measurement tools. Section 6 concludes the paper.

2   Timestamp Accuracy

We use the definition of accuracy found in [12], where accuracy is the comparison of a measured value and the correct value. The closer the measured value is to the correct the more accurate the measurement value is said to be. Each timestamp associated with a PDU has an accuracy denoted by T_Δ. A timestamp can either be associated with the arrival of the first bits or the last bits of the PDU, the method only requires that this assoiciation does not change, i.e. the timestamp cannot change association from the first to the last bit of the PDU during an evalution session. The timestamp accuracy is influenced by many sources, but primarily how often and how accurate a counter is updated as well as how fast it is possible to access this counter. In addition to these, the time- and frequency synchronization of the clock used for timestamping has a significant impact. Consider Fig. 1 where a PDU arrives at T_A, but due to the limited accuracy of the timestamp it is not possible to specify exactly when the packet actually arrived. The arrival can only be specified within an interval of with T_Δ, if the PDU arrived at T_A, then the value reported will be t_{n+2}. Please note that T_Δ contains all the error sources that can affect the timestamp, and at this stage there is no interest in differentiating them.

Now, depending on the measurement entity, the timestamp accuracy can vary significantly. For instance, in a time-shared environment (i. e., multi-tasking operating systems), the accuracy is not only affected by how often the counter is updated, but also of how the measurement entity can access this counter. Since the access usually involves a system call, there is an additional delay before the entity can register the arrival time. This delay is linked to the fact that there are multiple entities competing for the same resources (CPU, memory, etc.). From the accuracy point this reduces the accuracy of the measurement entity, see Fig. 2. When the GetTimestamp() call is made, there is a delay, $t_{request}$, before

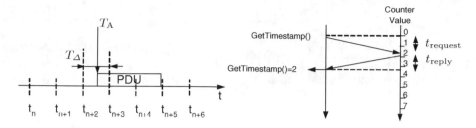

Fig. 1. Timestamp inaccuracy   **Fig. 2.** Obtaining a timestamp

the call reaches the counter. Once the execution reached the counter there is an another delay, t_{reply}, before the value, 2, is returned to the caller. Now if t_{request} and t_{reply} would be fixed in length, it would be possible to adjust for these delays, but in a multi-tasking environment they are usually not known. Thus, instead of the correct timestamp value 0, the value 2 would be reported.

3 Method

The principle in this method is to generate a traffic stream with a known and stable behaviour. As long as the stream is stable it can be generated at any layer. However the simplest and cheapest way is to use the link layer to generate identically sized PDUs that are sent back-to-back at the physical layer. The traffic stream is then monitored by the measurement system and inter-arrival times are calculated. By generating back-to-back PDUs at the physical layer, the impact of hardware and software in the generating system will be minimised. On the other hand, it relies on the correct implementation of the physical layer protocol in the hardware.

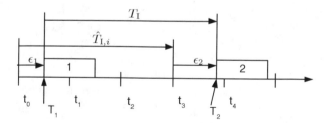

Fig. 3. PDU arrival

Let $\hat{t}_i$ be the reported arrival time of PDU i, T_i represent the *true* time and $\epsilon_i = T_i - \hat{t}_i$, i.e., the error for PDU i. This will assume values between 0 and T_Δ. In Fig. 3, PDU 1 arrives at T_1 and PDU 2 at T_2, PDU 1 gets the timestamp $\hat{t}_1 = t_0$ and PDU 2 the timestamp $\hat{t}_2 = t_3$. Based on this, we calculate an estimate of the inter-arrival time $\hat{T}_{I,i}$ of PDU i and $i + 1$:

$$\hat{T}_{I,i} = \hat{t}_{i+1} - \hat{t}_i = t_3 - t_0 = (T_{i+1} - \epsilon_{i+1}) - (T_i - \epsilon_i) = (T_{i+1} - T_i) - \epsilon_{i+1} + \epsilon_i. \quad (1)$$

For the example this will be $T_{I,1} = (T_2 - T_1) - \epsilon_2 + \epsilon_1$. The theoretical inter-arrival time, T_I, is calculated from the PDU length L at the physical layer (including inter-frame gap) and the link capacity C as $T_I = L/C$. If we subtract this from the measured inter-arrival time an estimate of the error is obtained:

$$\varepsilon_i = \hat{T}_{I,i} - T_I = (T_2 - T_1) - T_I + \epsilon_1 - \epsilon_2 = \epsilon_1 - \epsilon_2. \quad (2)$$

This error will assume value between $-T_\Delta$ and T_Δ, and it describes the combined error of two timestamps.

Let the theoretical inter-arrival time be written as:

$$T_I = nT_\Delta + \alpha T_\Delta \quad n = 0, 1, 2, \ldots \quad \alpha \in [0, 1[. \quad (3)$$

Here two cases can be found, in the first case $\alpha = 0$ and in the second $0 < \alpha < 1$. In the first case, the theoretical and estimated inter-arrival times will almost always be identical, and the calculated error will become:

$$\varepsilon_i = \hat{T}_{I,i} - T_I = nT_\Delta - nT_\Delta = 0. \quad (4)$$

However, due to numerical inaccuracies some samples will become either $-T_\Delta$ or $+T_\Delta$. For this to happen, the inter-arrival times must be a multiple of T_Δ. In case two, this is excluded by definition. Here the estimated inter-arrival $\hat{T}_{I,i}$ becomes:

$$\hat{T}_{I,i} = \begin{cases} nT_\Delta & \text{Counter not incremented (Case 2a)} \\ nT_\Delta + T_\Delta & \text{Counter incremented} \quad \text{(Case 2b)} \end{cases} \quad (5)$$

and ε_i becomes:

$$\varepsilon_i = \begin{cases} nT_\Delta - (nT_\Delta + \alpha T_\Delta) = -\alpha T_\Delta & \text{(Case 2a)} \\ nT_\Delta + T_\Delta - (nT_\Delta + \alpha T_\Delta) = (1 - \alpha)T_\Delta & \text{(Case 2b)} \end{cases} \quad (6)$$

effectively causing ε_i to alternate between $-\alpha T_\Delta$ and $-\alpha T_\Delta + T_\Delta$. This behaviour is easy to detect if a histogram of ε is plotted. The histograms can be classified as type 1 (Case 1) or type 2 (Case 2). A type 1 histogram has either one, at 0, or three peaks, at $-T_\Delta, 0$ and $+T_\Delta$. Fig. 4 shows a type 1 histogram, and a type 2 histogram is shown in Fig. 5. It will consist of only two peak values at $-\alpha T_\Delta$ and $-\alpha T_\Delta + T_\Delta$. Due to numerical reasons, a peak can cover two histogram bins. By analysing ε an estimate of T_Δ can be obtained from the extreme values:

$$T_\Delta = \begin{cases} \frac{|\max(\varepsilon)| + |\min(\varepsilon)|}{2} & \text{(Case 1)} \\ |\max(\varepsilon)| + |\min(\varepsilon)| & \text{(Case 2)} \end{cases} \quad (7)$$

One way to determine if a system shows a case 1 or 2 behaviour is to study ε in detail and to build a histogram. This results however in a more complex

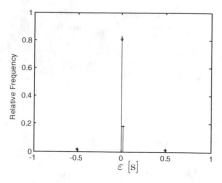

Fig. 4. Type 1 histogram of ε

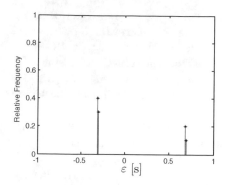

Fig. 5. Type 2 histogram of ε

implementation of the method. On the other hand, if the method always assumes that the histograms are of type 2, then the estimation simply becomes:

$$T_\Delta = |\max(\varepsilon)| + |\min(\varepsilon)|. \tag{8}$$

Using this approach, there is a chance that the accuracy might be underestimated by a factor of two. However, this is preferred compared to overestimating the accuracy.

There are also two practical problems associated with this method. If the traffic generator and the measurement system clocks are or become synchronised in such a way that ϵ_i will always be the same, this will result in a zero error, i.e. $\varepsilon_i = 0$. This has however never been detected in measurements as the crystal inaccuracies obviously work in our favour. Another problem is related to the quality of the traffic generator. At low speeds, smaller than 100 Mbps, it should not be any problem to generate PDUs back-to-back in a standard PC with a CPU of 2.0 GHz or more and Linux. However, at higher speeds this could be a critical issue. In this case, an alternative is to use a reference system, that has a known and high timestamp accuracy, to obtain an inter-arrival trace and use this instead of the theoretical inter-arrival time T_I, i.e., replace T_I in Equation 4 with the inter-arrival times calculated by the reference system.

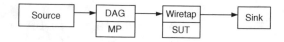

Fig. 6. Measurement setup

4 Setup

Using the method described in the previous section, we evaluated two hardware based measurement systems, the DAG 3.5E card and the Agilent J6800/6830A. Moreover, we evaluated the Packet Capture Library (PCAP) for three different operating systems (Linux 2.4, Linux 2.6 and FreeBSD 5.3) using different sets of hardware. We also evaluated a raw socket system. A raw socket allows an application to connect directly to a NIC. This allows the application to receive link layer frames from the NIC. Two versions of the raw socket interface were evaluated. The first one uses `ioctl(fd,SIOCGSTAMP,&tv)`, to read the kernel-associated timestamp with the PDU. This is the same method used by PCAP to obtain the PDU timestamp. The other one uses an assembler call to read the processor counter, known as the TSC method [8]. The difference consists in that `ioctl` should report the time when the kernel sees the PDU, while the TSC reports when the PDU is seen by the application, in this case the capture interface. The TSC method was evaluated on a P4-2.0 GHz system with Linux 2.4 and also on a P4-2.8 GHz system with Linux 2.6. To obtain the actual CPU speed an Acutime 2000 (Port B) was connected to the serial port of the PCs, and the number of CPU cycles that passed between the GPS pulses were counted, the result of which was then used to estimate the inter-arrival time. During these tests the CPU speed was estimated to 1 992 643 954 Hz (std.dev 700 Hz) for the Linux 2.4 PC (P4") and 2 800 232 626 Hz (std.dev 2 500 000 Hz) for the Linux 2.6 PC (P4').

All systems were evaluated on an 10 Mbps Ethernet and full size PDUs were used, resulting in a theoretical inter-arrival time of 1230400 ns (approximately 812 frames/s). All test were executed using the distributed passive measurement infrastructure presented in [11]. This consists out of measurement points (MP) that perform the packet capturing and consumers that analyse the measurement trace obtained from the MPs. The MPs use capture interfaces to attach to the monitored network. These can be DAG cards or software. The setup is shown in Fig. 6. The traffic was generated by Source, using the Linux Packet Generator kernel module on Pentium-4 processor of 2.8 GHz with Linux 2.6 and 1024 MB of RAM. In [1] we show that the generator is stable, i.e., generates PDUs back-to-back, under these conditions. The traffic was then fed to a reference measurement point that used a DAG 3.5E capture interface, providing the facility to validate the correct behaviour of the traffic generator during the tests. A wiretap split the signal so that the system under test (SUT) obtained a copy, while the original traffic stream was terminated in the Sink. The SUT was implemented as a capture interface in the measurement point, with the exception of the Agilent system that uses a custom built PC. Here, we wrote a script that converted the logs provided by the Agilent software into a format usable by our analysis software.

The hardware systems were evaluated using 100 000 PDUs, while the software used both 100 000 and 250 000 PDUs. Furthermore, PCAP, PF_RING, MMAP and `ioctl` tests were executed at different times and on two different hardware platforms, a Pentium-3 664 MHz and a Pentium-4 2.4 GHz. The clocks of the MPs were synchronised using NTP. However, as we only show the evaluation over a short time interval the impact of NTP is not obvious. If NTP would have had to make a time correction this would have been clearly notable as one (or more) inter-arrival times would have been much larger, or negative. The TSC test was executed on two different Pentium-4's as described above.

5   Results

In this configuration the DAG 3.5E card obtained a timestamp accuracy of 59.75 ns, which is equivalent to the card's clock resolution. The Agilent system obtained a timestamp accuracy of 100 ns, which matches the reported resolution of its timestamp. In Fig. 7 the error histogram for the DAG 3.5E is shown, and Fig. 8 contains the histogram for the Agilent system. In both cases the bin width was 1 ns and the histograms are of type 1.

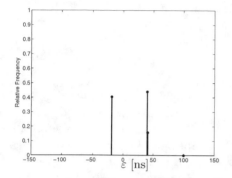
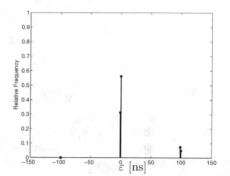

Fig. 7. Error histogram for DAG 3.5E, bin width 1 ns

Fig. 8. Error histogram for Agilent J6800/J6830A, bin width 1 ns

In Table 1 the estimated accuracies for PCAP, PF_RING and MMAP are shown. We observe that there is a large difference between the cases with 100 000 and the 250 000 PDU. Now, since NTP is employed even longer tests should be considered. But, as the test lengths grow, the synchronisation events will become very visible and will only demonstrate how good or bad the clock is in the particular computer. NTP conditions the clock both in time and frequency, and the time synchronisation can manifest itself with large negative, or positive, inter-arrival times. A 250 000 PDU test takes approximately five minutes to execute and as such it can act as a crude indicator of a system's accuracy, without too much interference from the clock synchronisation effects. Now, looking at PCAP one can see that there is not a very large difference between the operating systems for the 100 000 PDU test. In the 250 000 PDU test, FreeBSD performs significantly worse, with T_Δ estimations around 3 ms. It is also interesting to see that for the 100 000 PDU test, the P3 outperforms the P4 with T_Δ estimations around 0.2 ms regardless of the operating system. Turning the attention to PF_RING, one can see that it seems to require more processing than PCAP for the P3, resulting in a worse T_Δ value. For the P4, the accuracy relative to PCAP increases slightly for Linux 2.4, but decreases for Linux 2.6. Now, looking at MMAP for 100 000 PDUs, the P3 behaves similarly as in the PCAP case. However, the P4 is much better with a T_Δ of 0.023 ms for Linux 2.4 and 0.14 ms for Linux 2.6. This indicates a quite efficient system. But, when increasing the test length to 250 000 PDUs, it obtains roughly the same accuracy as PCAP.

Table 1. T_Δ estimations for PCAP, PF_RING and MMAP

Measurement Software	Operating System	T_Δ [μs]			
		100 000 PDUs		250 000 PDUs	
		P3-664 MHz	P4-2.4 GHz	P3-664 MHz	P4-2.4 GHz
PCAP	Linux 2.4.29	202	415	720	346
	Linux 2.6.10	218	375	663	374
	FreeBSD 5.3	229	375	2 760	3 243
PFRING	Linux 2.4	746	296	810	320
	Linux 2.6	730	480	810	440
MMAP	Linux 2.4	489	23	540	340
	Linux 2.6	300	137	700	460

In Table 2 estimations of T_Δ are presented for the raw socket. The P3 exhibits an interesting behaviour: When changing from Linux 2.4 to 2.6 with NTP, it almost doubles its accuracy. Overall the accuracy ranges from 0.3 ms for the P4 with Linux 2.4 to 0.71 ms for the P3 and Linux 2.4. In the histograms, the two peaks are visible at the same places and the shapes are similar to the ones seen for the PCAP, PF_RING and MMAP systems. For the TSC method, for Linux 2.6 the 2.8 GHz obtained the same result as the 2.4 GHz for the `ioctl` case. But for the Linux 2.4, the TSC method was worse than the 2.4 GHz `ioctl` case.

Table 2. T_Δ estimations for raw socket

Operating System	T_Δ [μs]			
	ioctl		TSC	
	P3-664 MHz	P4-2.4 GHz	P4-2.0 GHz	P4-2.8 GHz
Linux 2.4	710	300	574	N/A
Linux 2.6	330	410	N/A	410

In Fig. 9 the error histograms for the five software measurement systems are shown, based on the 250 000 PDU test. The histograms have bins that are 10 μs wide. First, one can see the similarity between all five systems. Second, the widths of the histograms are in the same order. All five histograms consist of five regions. The two peaks that stick out are at $[-110, -100[$ μs and $[10, 20[$ μs. The left most peak holds approximately 10 % of the samples, while the other holds around 80 % of the samples. A probable cause for the peaks is the scheduling by the operating system. When a detailed analysis is done on the first 100 PDUs captured by the P3 using Linux (both 2.4 and 2.6), approximately six out of seven ε samples are around 20 μs, while the seventh sample has an ε of approximately -110 μs. This is clearly a type 2 behaviour, indicating that T_Δ should be 130 μs. However, as the trace grows more 'jitter' appears, resulting in T_Δ estimations up to six times larger. Now, this jitter is a part of the system and can as such not be ignored. Hence, its influence must be included into the accuracy estimation of T_Δ.

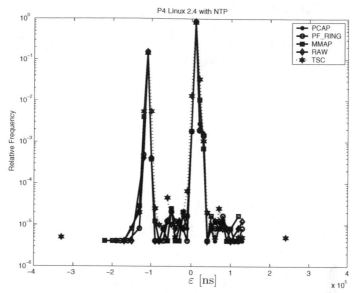

Fig. 9. Error histograms for Linux 2.4 based systems, bin width 10 μs

Another observation is that since all systems have the same peak, they are all subject to the same scheduling. Now, since the TSC method obtains its timestamp at the application level, it should be subject to operating system scheduling with *user* priority. Now consider the `ioctl` method that is supposed to read the timestamp associated with the PDU. This timestamp is assumed to be set by the kernel, however it shows the same behaviour as the TSC method. Hence it must be also subject to user scheduling, and not kernel scheduling as it should. This implies that the `ioctl` method obtains the timestamp at the application level under user scheduling constraints. And, since all the other three systems also show the same behaviour, they most probably collect their timestamps at the application level as well. Hence, the timestamps does not only reflect what happens on the network but also what happens in the computer.

6   Conclusions

In this paper we presented a promising method to estimate the timestamp accuracy obtained from measurement equipment, both hardware and software. Knowing the timestamp accuracy of a measurement system is the first step in figuring out the quality of the performance parameters calculated from traces obtained from that particular system.

Using the method, we evaluated the DAG 3.5E and the Agilent J6800/J6830A and were able to confirm that the DAG has a timestamp accuracy of 60 ns and the Agilent has an accuracy of 100 ns. We also evaluated five software tools for various software and hardware configurations. With regards to the software tools tested here, two conclusions can be drawn. First, they timestamp PDUs at the application level instead of kernel level. Hence, the timestamp does not match the data delivered to the measurement tool. Second, the estimated timestamp accuracy T_Δ is in the same order of magnitude, around 0.3 ms to 0.8 ms using Linux and around 3 ms using FreeBSD together with off-the-shelf components and software.

To find the timestamp accuracy, it is obvious that long evaluations are needed in order to correctly capture the underlying distribution. However, as the evaluation time grows, clock synchronization starts to influence the results significantly. The best approach is to evaluate a system over a time frame that covers the system's normal operating hours. This way the worst case accuracy is known before measurements are performed, and thus, the measurement setup can be calibrated by choosing components that allow for the desired precision.

Future work will include the construction of a traffic generator, implemented in VHDL, to allow the generation of highly accurate PDU inter-arrival times.

As of now, current best practice is to evaluate the timestamp accuracy of a measurement system prior to use, and if needed change hardware, operating system and configuration, to obtain an acceptable timestamp accuracy.

References

1. P. Arlos: On the Quality of Computer Network Measurements, Ph.D. Thesis, Blekinge Institute of Technology, 2005.
2. A. A. Ali, F. Michaut and F. Lepage: End-to-end Availible Bandwidth Measurement Tools: A Comparative Evaluation of Performance, Proc. 4th International Workshop on Internet Performance, Simulation, Monitoring and Measurement, 2006.
3. Endace Measurement Systems: http://www.endace.com, verif. Jan/2007.
4. S. Donnelly: High Precision Timeing in Passive Measurements of Data Networks, Ph.D. Thesis, The University of Waikato, 2002.
5. TCPDUMP: Homepage. http://www.tcpdump.org, verif. Jan/2007.
6. L. Deri: PF_RING: Passive Packet Capture on Linux at High Speeds, http://www.ntop.org/PF_RING.html, verified in Jan/2007.
7. P. Wood: Memory Mapping for PCAP, http://public.lanl.gov/cpw/, verif. Jan/2007.
8. D. Veitch and S. Babu and A. Pásztor: Robust Synchronization of Software Clocks Across the Internet, Proc. Internet Measurement Conference, 2004.
9. P. Arlos and M. Fiedler: A comparison of Measurement Accuracy for DAG, Tcpdump and Windump. http://www.its.bth.se/staff/pca/aCMA.pdf, verif. Jan/2007.
10. WINDUMP: Homepage. http://www.winpcap.org/windump/, verif. Oct/2006.
11. P. Arlos, M. Fiedler and A. Nilsson: A Distributed Passive Measurement Infrastructure, Proc. Passive and Active Measurement Workshop, 2005.
12. D. Buchla and W. McLachlan: Applied Electronic Instrumentation and Measurement, ISBN 0-675-21162-X, 1992.

Packet Capture in 10-Gigabit Ethernet Environments Using Contemporary Commodity Hardware

Fabian Schneider, Jörg Wallerich, and Anja Feldmann

Deutsche Telekom Laboratories / Technische Universität Berlin
10587 Berlin, Germany
{fabian,joerg,anja}@net.t-labs.tu-berlin.de
http://www.net.t-labs.tu-berlin.de

Abstract. Tracing traffic using commodity hardware in contemporary high-speed access or aggregation networks such as 10-Gigabit Ethernet is an increasingly common yet challenging task. In this paper we investigate if today's commodity hardware and software is in principle able to capture traffic from a fully loaded Ethernet. We find that this is only possible for data rates up to 1 Gigabit/s without reverting to using special hardware due to, e. g., limitations with the current PC buses. Therefore, we propose a novel way for monitoring higher speed interfaces (e. g., 10-Gigabit) by distributing their traffic across a set of lower speed interfaces (e. g., 1-Gigabit).

This opens the next question: which system configuration is capable of monitoring one such 1-Gigabit/s interface? To answer this question we present a methodology for evaluating the performance impact of different system components including different CPU architectures and different operating system. Our results indicate that the combination of AMD Opteron with FreeBSD outperforms all others, independently of running in single- or multi-processor mode. Moreover, the impact of packet filtering, running multiple capturing applications, adding per packet analysis load, saving the captured packets to disk, and using 64-bit OSes is investigated.

Keywords: Packet Capturing, Measurement, Performance, Operating Systems.

1 Introduction

A crucial component of almost any network measurement and especially any network security system is the one that captures the network traffic. For example, nowadays almost all organizations secure their incoming/outgoing Internet connections with a security system. As the speed of these Internet connection increases from T3 to 100-Megabit to 1-Gigabit to 10-Gigabit Ethernet, the demands on the monitoring system increase as well while the price pressure remains. For example, the Münchner Wissenschaftsnetz (MWN, Munich Scientific Network) [1] offers Internet connection to roughly 50,000 hosts via a 10-Gigabit bidirectional uplink to the German Scientific Network (DFN). While the link is currently rate limited to 1.2-Gigabit we indeed face the challenge of performing packet capture in a 10-Gigabit Ethernet environment using commodity hardware in order to run a security system. We note that most network security systems

S. Uhlig, K. Papagiannaki, and O. Bonaventure (Eds.): PAM 2007, LNCS 4427, pp. 207–217, 2007.

rely on capturing *all* packets as any lost packet may have been the attack. Furthermore, most attack detection mechanisms rely on analyzing the packet content and the security system itself has to be resilient against attacks [2]. Therefore packet capture has to be able to handle both the maximum data rates as well as the maximum packet rates.

Unfortunately, network traffic capture is not an easy task due to the inherent system limitations (memory and system bus throughput) of current off-the-shelf hardware. One expensive alternative[1] is to use specialized hardware, e. g., the monitoring cards made by Endace [3]. Our experience shows that PCs equipped with such cards are able to capture full packet traces to disk in 1-Gigabit environments. But even these systems reach their limits in 10-Gigabit environments: they lack CPU cycles to perform the desired analysis and/or bus bandwidth to transfer the data to disk. Therefore, we in this paper propose a novel way for distributing traffic of a high-speed interface, e. g., 10-Gigabit, across multiple lower speed interfaces, e. g., 1-Gigabit, using of-the-shelf Ethernet switches. This enables us to use multiple machines to overcome the system limitations of each individual interface without losing traffic.

This leaves us with the question, which system is able to monitor one such 1-Gigabit Ethernet interface given the many factors that impact the performance of a packet capture system. These include the system, CPU, disk and interrupt handling architecture [4], the operating system and its parameters, the architecture of the packet capture library [5,6,7,8], and the application processing the data [2,5,9]. In this paper, we focus on the first set of factors and only consider simple but typical application level processing, such as compressing the data and storing it to disk. We chose to examine three high-end off-the-shelf hardware architectures (all dual processor): Intel Xeon, AMD Opteron single core and AMD Opteron multi-core based systems; two operating systems: FreeBSD and Linux with different parameter settings; and the packet capture library, libpcap [5], with different optimizations.

In 1997, Mogul et al. [4] pointed out the problem of receive livelock. Since then quite a number of approach [10,11,7,8] to circumvent this problem have been proposed. Yet the capabilities and limitations of the current capturing systems have not been examined in the recent past.

The remainder of this paper is structured as follows. In Sec. 2 we discuss why capturing traffic from an 10-Gigabit Ethernet link is nontrivial and propose ways of distributing traffic across several 1-Gigabit Ethernet links. In Sec. 3 we ask if it is feasible to capture traffic across 1-Gigabit links using commodity hardware. Next, in Sec. 4, we discuss our measurement methodology and setup. Sec. 5 presents the results of our evaluation. Finally, we summarize our contributions and discuss future work in Sec. 6.

2 Challenges That Hinder Data Capture at High Rates

With current PC architectures there are two fundamental bottlenecks: bus bandwidth and disk throughput. In order to capture the data of a fully utilized bidirectional 10-Gigabit link one would need 2560 Mbytes/s system throughput. Even, for capturing only the packet headers, e. g., the first 68 bytes, one needs 270 Mbytes/s given an observed average packet size of 645 bytes. Moreover, when writing to disk, packet capture

---

[1] Current cost for a 1-Gigabit card is roughly 5,000 €.

libraries require the data to pass the system bus twice: Once to copy the data from the card to the memory to make it accessible to the CPU, and once from memory to disk. Thus this requires twice the bandwidth.

When comparing these bandwidth needs with the bandwidth that current busses offer we notice a huge gap, especially for full packet capture. The standard 32-bit, 66 MHz PCI bus offers 264 Mbytes/s. The 64-bit, 133 MHz PCI-X bus offers 1,066 Mbytes/s. The PCIexpress x1 bus offers 250 Mbytes/s. It is easy to see that none of these busses are able to handle the load imposed by even a single uni-directional 10-Gigabit link (full packets). Furthermore the numbers have to be taken with a grain of salt as they are maximum transfer rates, which in case of PCI and PCI-X are shared between the attached PCI cards. Some relief is in sight, PCIexpress x16 (or x32) busses which offer 4000 Mbytes/s (8000 Mbytes/s) are gaining importance. Currently boards with one or two x16 slot are available. Yet, note that these busses are intended for graphic cards and that 10-Gigabit Ethernet cards are only available for the PCI-X bus at the moment. Indeed, at this point we do not know of any network or disk controller cards that support PCIexpress x16 or x32.

But what about the bandwidth of the disk systems. The fastest ATA/ATAPI interfaces runs at a mere 133 Mbytes/s; S-ATA (2) offers throughputs of up to 300 Mbytes/s; SCSI (320) can achieve 320 Mbytes/s. Even the throughput of Fiberchannel, up to 512 Mbytes/s, and Serial-Attached-SCSI (SAS), up to 384 Mbytes/s, is not sufficient for 10-Gigabit links. The upcoming SAS 2 systems are going to offer 768 Mbytes/s while SAS 3 systems may eventually offer 1536 Mbytes/s. Again there is a huge gap.

Therefore, it appears unavoidable to distribute the load across multiple machines. Instead of designing custom hardware for this purpose we propose to use a feature of current Ethernet switches. Most switches are capable of bundling a number of lower speed interfaces. Once the bundle has been configured it can be used like a normal interface. Therefore it can be used as a monitor interface for any other interface. Accordingly, the switch forwards all traffic that it receives on a, e. g., 10-Gigabit interface on a bundle of, e. g., 1-Gigabit interfaces. These lower speed interfaces can then be monitored individually. Fig. 1 shows a schematic view of this setup. There are three major drawbacks to this solution: first, time stamps are slightly distorted as the packets have to pass through an additional switch; second, it is hard to synchronize the system time of the monitors; third, even though the overall capacity suffices the individual capacity of each link may not be sufficient. This depends on the load balancing scheme in use.

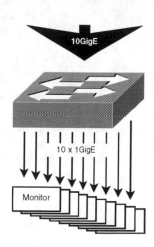

Fig. 1. Distr. Setup

We tested this mechanism by using a Cisco 3750 switch to move from monitoring a single 1-Gigabit Ethernet link to eight (which is maximum for this vendor) 100-Mbit links. We use the EtherChannel feature and configured such a link bundle across eight 100-Mbit links. Then this link is used as a monitor link for a 1-Gigabit input link.

It is crucial to use the appropriate load balancing method for this link. Common options include load balancing based on MAC addresses (simple switches), IP addresses and/or MAC addresses (e. g., Cisco 3750), combinations of IP addresses and/or port (TCP/UDP) numbers etc. (Cisco 6500 Series). Note that per packet multiplexing is not a sensible option as this can lead to an uneven distribution across the monitors. Given that it is common to use a switch for monitoring traffic traveling between two routers the MAC address variability is limited. There are usually only two MAC addresses in use, one for each endpoint. Accordingly, we have to rely on at least source and destination IP addresses or better yet IP addresses and port numbers for load balancing which unfortunately rules out the cheapest switches. Still, depending on the application one may want to ensure that all packets from an IP address pair or a remote or local host are handled by a single monitor. This ensures that all packets of a single TCP connection arrive at the same monitor. With such a configuration load balancing can be expected to be reasonable as long as there are sufficiently many addresses that are monitored. If the EtherChannel feature is not supported by a specific switch one should check if it offers another way to bundle links. A good indication that a switch is capable of bundling is the support of the Link Aggregation Control Protocol (LACP).

3   Influencing Parameters and Variables

Given that we now understand how we can distribute the traffic from an 10-Gigbit interfaces across multiple 1 Gigabit interfaces we next turn our attention to identify suitable commodity hard- and software for the task of packet capturing. For this purpose we identify the principal factors that may impact the performance and then examine the most critical ones in detail.

Obvious factors include the network card, the memory speed, and the CPU cycle rate. A slightly less obvious one is the system architecture: does it matter if the system has multiple CPUs, multiple cores, or if hyper-threading is enabled? Another important question is how the OS interacts with the network card and passes the captured packets from the card to the application. How much data has to be copied in which fashion? How many interrupts are generated and how are they handled? How much buffer is available to avoid packet loss? Yet another question is if a 64-bit OS offers an advantage over the 32-bit versions. With regards to the application we ask by how much the performance is degraded when running multiple capturing applications on the same network card, when adding packet filters, when touching the data in user space, i. e., via a copy operation or by compressing them, and when saving the data to a tracefile on disk. These are the questions that we try to answer in the remainder of the paper.

4   Setup

Conducting such measurements may appear simple at first glance, but the major difficulty is to find hard-/software configurations that are comparable. After all we want to avoid comparing apples to oranges.

4.1 Hardware and Software

To enable a fair comparison, state-of-the-art off-the-shelf computer hardware was purchased at the same time, February 2004, with comparable components. Preliminary experiments have shown that the Intel Gigabit Ethernet cards provide superior results than, e. g., those of Netgear. Furthermore, 2 Gbytes of RAM suffice and we choose to use the fastest available ones. Accordingly, we focus on the system architecture and the OS while using the same system boards, the same amount of memory, and the disk system. Consequently, we purchased two AMD Opteron and two Intel Xeon systems[2] that are equipped with the same optical network card, the same amount of memory, and the same disk system (IDE ATA based). One Opteron and one Xeon system were installed with FreeBSD 5.4 and the others were installed with Debian Linux (Kernel v2.6.11.x).

Once dual-core systems became available we got two additional machines in May 2006: HP Proliant DL385[3]. In contrast to the other systems these have an internal SCSI Controller with three SCSI disk attached. One was installed with FreeBSD 6.1 and the other with Debian Linux (Kernel v2.6.16.16), both with dual-boot for 32-bit and 64-bit.

4.2 Measurement Topology

To be able to test multiple systems at the same time and under exactly the same workload we choose the setup shown in Fig. 2. A workload generator, see Sec. 4.4, generates traffic which is fed into a passive optical splitter. The task of the splitter is to duplicate the traffic such that all systems under test (SUT) receive the same input. The switch between the workload generator and the splitter is used to check the number of generated packets while all

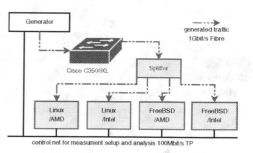

Fig. 2. Measurement Topology

capture applications running on the SUT's are responsible for reporting their respective capture rate. In later experiments we removed the switch as the statistics reported by the traffic generator itself proved to be sufficient. The control network is a separate network and is used to start/end the experiments and to collect the statistics.

4.3 Methodology

Each measurement varies the traffic rate from 50 Mbit/s to 920 Mbit/s[4] and consists of seven repetitions of sending one million packets at each bandwidth setting to reduce

---

[2] System details: AMD Opteron 244, Intel Xeon 3.06Ghz, 2 Gbytes RAM (DDR-1 333 MHz), Intel 82544EI Gigabit (Fiber) Ethernet Controller, 3ware Inc. 7000 series ATA-100 Storage RAID-Controller with at least 450 Gbytes space.

[3] Details see previous footnote except of processors (AMD Opteron 277), disk system (HP Smart Array 64xx Controller with 3 × 300 Gbytes U320 SCSI hard disks), and faster memory.

[4] 920Mbit/s is close to the maximum achievable theoretical input load given the per packet header overheads.

statistical variations. In each iteration we start the capture process (using a simple libp-cap [5] based application [12] or tcpdump) and a CPU usage profiling application on the systems under study. Then we start the traffic generator. Once all packets of the run have been sent by the workload generator we stop the packet capture processes.

For each run we determine the capture rate, by counting the received number of packets, and the average CPU usage (see cpusage [12]). The results are then used to determine the mean capturing rate as well as the mean CPU usage for each data rate.

4.4   Workload

With regards to the generated traffic we decided to neither focus on the simplest case for packet monitoring (all large packets) nor on the most difficult one (all small packets) as both are unrealistic. The first case puts too much weight on the system bus performance while the latter one emphasizes the interrupt load too much.

Instead we decided to use a traffic generator that is able to generate a packet size distribution that closely resembles those observed in actual high-speed access networks. For this purpose we enhanced the Linux Kernel Packet Generator (LKPG [13]), which is capable of filling a 1-Gigabit link, to generate packets according to a given packet size distribution.

Figure 3 shows the input and the gener-ated packet size distri-butions of our modified LKPG based on a 24h packet level trace cap-tured at the 1-Gigabit uplink of the MWN [1] (peaks are at 40–64, 576 and 1420–1500 B).

We decided to not mimic flow size dis-tribution, application layer protocol mix, de-lay and jitter, or data content, as they have no direct impact on the performance of the capture system. Their impact is on the appli-cation which is beyond the scope of this paper. The same holds for throughput bursts, which can be buffered. The intention of the work is to identify the maximum throughput a capturing system can handle. We realize different traffic rates by inserting appropriate inter-packet gaps between the generated packets.

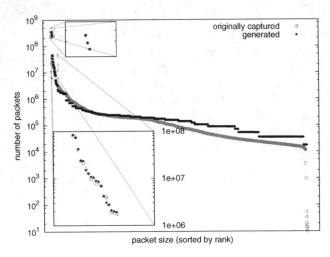

Fig. 3. Packet sizes (sorted by rank) vs. frequency of the packets (y-axis in logscale)

5 Results

We now use the above setup to evaluate the performance of our systems starting from the vanilla kernels. We find that it is crucial to increase the amount of memory that is available to the kernel capture engine. Otherwise scheduling delays, e. g., to the capturing application, can cause the overflow of any of the kernel queues, e. g., the network queue for the capture socket. This is achieved by either setting the `debug.bpf_bufsize` sysctl parameter (FreeBSD) or the `/proc/sys/net/core/rmem*` parameter (Linux). Based upon our experience we chose to use buffers of 20 Mbytes. Furthermore, we noticed that the systems spend most of their cycles on interrupt processing when capturing at high packet rates. To overcome this problem we tried device polling as suggested by Mogul et al. [4]. For Linux this reduces the CPU cycles that are spent in kernel mode and thus increases the packet capture rate significantly. For FreeBSD activating the polling mechanism slightly reduced the capturing performance and the stability of the system. Therefore we use it for Linux but not for FreeBSD.

Next we baseline the impact of the system architecture and the OS by comparing the performance of the four systems with single processor kernels. Figure 4 (top) shows the capture rate while Fig. 4 (bottom) shows the CPU usage as we increase the data rate. To keep the plots simple we chose to not include the standard deviation. Note that all measurements have a standard deviation of less than 2%. As expected the capture rate decreases while the CPU usage increases as the data rate is increased. Unfortunately, only the FreeBSD/AMD combination loses almost no packets. All others experience significant packet drops. We note that the systems start dropping packets once their CPU utilization reaches its limits. This indicates that the drops are not due to missing kernel buffers but are indeed due to missing CPU cycles. The FreeBSD/Intel combination already starts dropping packets at about 500 Mbit/s. Neither of the Linux systems looses packets until roughly 650 Mbit/s. From that point onward their performance deteriorates

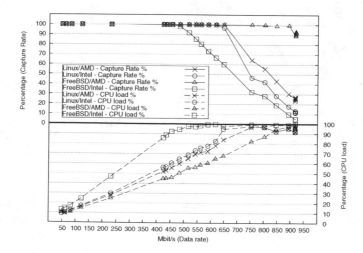

Fig. 4. Data rate vs. Capture Rate (top) and CPU utilization (bottom) for: single processor; increased buffers

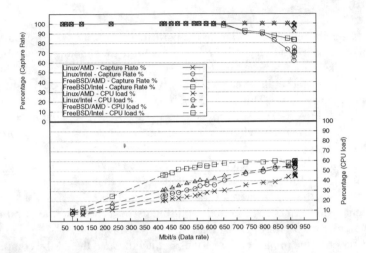

Fig. 5. Data rate vs. Capture Rate (top) and CPU utilization (bottom) for: multiple processors; increased buffers. (A CPU usage of 50% implies that one CPU is fully utilized.)

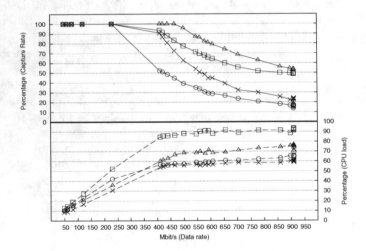

Fig. 6. Data rate vs. Capture Rate (top) and CPU utilization (bottom) for: multiple processors; increased buffers; 50 additional `memcpy` operations on the packet data. (To increase readability the legend for this plot is omitted as it is the same as in Fig. 4.)

dramatically with increasing data rates. The efficiency of the FreeBSD/AMD machine is especially surprising as FreeBSD performs an additional (kernel) packet copy operation and does not use device polling which proved beneficial on Linux systems.

This observation indicates that utilizing the multi-processor system may help over-come the above problems as the kernel can be scheduled on one processor while the application is scheduled on the other. This almost doubles the CPU resources. Yet,

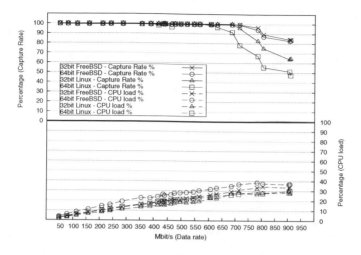

Fig. 7. Data rate vs. Capture Rate (top) and CPU utilization (bottom) for: multiple *dual-core* processors; increased buffers; writing full packets to disk. (A CPU usage of 25% implies that one CPU is fully utilized.)

memory conflicts and cache misses have the potential to deteriorate the capture performance. The results shown in Fig. 5 show the resulting significant performance increase. This indicates that the additional CPU cycles of the SMP architecture clearly top their penalties, e. g., cache misses. In fact, the FreeBSD/AMD combination is able to capture all packets even at the highest data rate. Overall we note that in our experience FreeBSD outperforms Linux. To test if further increasing the multiprocessor capabilities helps the performance we enabled Hyper-Threading (only available on Intel machines). This does not impact the performance. But keep in mind that in this test only two processes require significant CPU resources, the kernel and the capture application.

As it is common to use filters while capturing, we studied the impact of configuring a 50 BFP instructions filter. We find that even if packets have to pass a long filter before being accepted that does not drastically change the performance. The performance of the FreeBSD machines stays as is while the one of the Linux machines decreases slightly (up to 10%) at high data rates (>800 Mbits/s). This indicates that at least for FreeBSD filtering does not impose high costs on CPU resources. But what if we run multiple capture applications at the same time? In this case the packets have to be duplicated within the kernel and delivered to multiple filter instances. We not surprisingly find that the performance deteriorates for all four systems. As the Linux machines start to massively drop packets when reaching their CPU limits, we suggest to use FreeBSD as the drop rates are less significant.

The next challenge that we add to the application are memory copy operations as these are common in network security applications. To investigate an extreme position we instructed the application to copy the captured packet 50 times in user-space. From Fig. 6 we see that all systems suffer under the additional load. The Linux machines yet again experience larger problems. The next challenge is to compress the content of each packet using a standard compression routine, `libz`. For the first time the Intel systems outperform their AMD counterparts. This suggests that Intel processors and not AMD's have better hardware realizations for instructions used by `libz`.

Recapitulating, we find that the multiprocessor Opteron system with FreeBSD 5.4 outperforms all other systems. Its maximum loss rate is less than 0.5%. One explanation is that AMD has a superior memory management and bus contention handling mechanism. These results motivated us to only purchase Opteron dual-core machines for the next tests. To baseline the new system we repeated the above experiments, except the ones with additional application load. These experiments show that both machines can capture every single packet on FreeBSD as well as Linux running either the 32-bit or the 64-bit version of the respective OS.

The newer machines clearly outperformed the older ones. Therefore we next examine if either system is able to capture full packet traces to disk. For this we switch to using `tcpdump` as application. From Fig. 7 we see that Linux is able to write all packets to disk up to a data rate of about 600 Mbit/s independent of using a 32-bit or 64-bit kernel. FreeBSD is always dropping roughly 0.1% of the packets even at the lowest data rates. This indicates a minor but fixable principle problem in the OS. FreeBSD only begins to drop a significant number of packets when the data rate exceeds 800 Mbit/s. But keep in mind that Linux captures only about 65% of the packets at the highest data rate (32-bit). While under FreeBSD the difference between 32-bit and 64-bit mode is negligible (up to a capture rate of 83%), Linux in 64-bit mode deteriorates drastically. It records only half of the generated packets. The poor performance of 64-bit Linux might be due to the increased memory consumption for longer pointers within the kernel.

6 Summary

In this paper we argue that due to currently available bus and disk bandwidth it is impossible to tackle the crucial problem of packet capture in a 10-Gigabit Ethernet environment using a single commodity system. Therefore we propose a novel way for distributing traffic across a set of lower speed interface using a switch feature that allows one to bundle lower speed interface into a single higher speed interface, e. g., Cisco's EtherChannels feature. Each of the lower speed interfaces can then be monitored using commodity hardware.

To answer the question which system is able to support packet monitoring best we present a methodology for evaluating the performance impact of various system components. We find that AMD Opteron systems outperform Intel Xeon ones and that FreeBSD outperforms Linux. While multi-processor systems offer a lot, the benefit of adding Hyper-Threading and multi-core is small. Moreover, it appears that for the task of packet capture the 64-bit OS versions are not quite ready yet. The newer systems clearly outperform the older ones and can even capture full packet traces to disk as long as the data rate is less than 600 to 700 Mbit/s. All in all, multi-processor Opteron systems with FreeBSD clearly outperformed all other systems.

Obviously, our goal has to be to understand not just the packet capture performance but the characteristics of such highly complex security screening applications as Bro.

Therefore we plan to investigate how system performance scales with the traffic load and if there is a way to predict future performance. A summary of our current and future results is available on the project Website [14].

References

1. The Munich Scientific Network. http://www.lrz-muenchen.de/wir/intro/en/#mwn
2. Paxson, V.: Bro: A System for Detecting Network Intruders in Real-Time. Computer Networks, **31(23-24)** (1999) 2435–2463
3. Endace Measurement systems: http://www.endace.com
4. Mogul, J. C., and Ramakrishnan, K. K.: Eliminating receive livelock in an interrupt-driven kernel. ACM Transactions on Computer Systems, **15(3)** (1997) 217–252.
5. Jacobson, V., Leres, C., and McCanne, S.: libpcap and tcpdump. http://www.tcpdump.org
6. Wood, P.: libpcap MMAP mode on linux. http://public.lanl.gov/cpw/
7. Deri, L.: Improving passive packet capture: Beyond device polling. In *Proc. of the 4th Int. System Administration and Network Engineering Conference (SANE'2004)* (2004)
8. Deri, L.: nCap: Wire-speed packet capture and transmission. In *Proc. of the IEEE/IFIP Workshop on End-to-End Monitoring Techniques and Services (IM 2005, E2EMON)* (2005)
9. Snort http://www.snort.org/
10. Salim, H. D., Olsson, R., and Kuznetsov, A.: Beyond softnet. In *Proc. of the 5th Annual Linux Showcase & Conference* (2001)
11. Rizzo, L.: Device Polling support for FreeBSD. In *Proc. of the EuroBSDCon' 01* (2001)
12. Schneider, F.: Performance Evaluation of Packet Capturing Systems for High-Speed Networks Diploma thesis, Technische Universität München (2005) for cpusage and the capturing application see: http://www.net.in.tum.de/~schneifa/proj_en.html
13. Olsson, R.: Linux kernel packet generator.
14. Hints for improving Packet Capture System performance: http://www.net.t-labs.tu-berlin.de/research/bpcs/

Neuro-fuzzy Processing of Packet Dispersion Traces for Highly Variable Cross-Traffic Estimation

Marco A. Alzate[1,2], Néstor M. Peña[1], and Miguel A. Labrador[3]

[1] Universidad de los Andes, Bogotá, Colombia
[2] Universidad Distrital, Bogotá, Colombia
[3] University of South Florida, Tampa, FL, USA
{m-alzate,npena}@uniandes.edu.co, labrador@cse.usf.edu

Abstract. Cross-traffic data rate over the tight link of a path can be estimated using different active probing packet dispersion techniques. Many of these techniques send large amounts of probing traffic but use just a tiny fraction of the measurements to estimate the long-run cross-traffic average. In this paper, we are interested in short-term cross-traffic estimation using bandwidth efficient techniques when the cross-traffic exhibits high variability. High variability increases the cross-correlation coefficient between cross-traffic and dispersion measurements on a wide range of utilization factors and over a long range of measurement time scales. This correlation is exploited with an appropriate statistical inference procedure based on a simple heuristically modified neuro-fuzzy estimator that achieves high accuracy, low computational cost, and very low transmission overhead. The design process led to a very simple architecture, ensuring good generalization properties. Simulation experiments show that, if the variability comes from a complex correlation structure, a single estimator can be used over a long range of utilization factors and measurement periods with no additional training.

Keywords: Traffic estimation, Packet pair dispersion, Neuro-fuzzy systems.

1 Introduction

Several network parameters and traffic conditions can be inferred from packet dispersion measurements, when a sender transmits probing packets of given length at given instants of time, and a receiver collects them taking note of their inter-arrival times [1]. For example, if the tight link is 100% busy between a pair of probing packets, the correlation coefficient between the dispersion measurements and the tight-link cross-traffic will be 1, i.e., the dispersion measurement reveals the average cross-traffic rate over that link [2]. Otherwise, this correlation will be less than one, increasing directly with the utilization factor and inversely with the probing packets inter-departure times.

Most available bandwidth estimation techniques send large amounts of probing traffic (overhead) in order to select the tiny fraction of measurements that satisfies the high correlation condition [3][4][5][6]. However, several simulation experiments with different synthetic traces exhibiting different degrees of variability (not shown here)

S. Uhlig, K. Papagiannaki, and O. Bonaventure (Eds.): PAM 2007, LNCS 4427, pp. 218–222, 2007.

reveal that this correlation can still be high over a wide range of link utilizations and over a long range of measurement time scales if the traffic's coefficient of variation is high and the traffic exhibits long range dependence. In this work, we consider a computational intelligence approach to estimate the competing traffic rate in the tight link that, instead of ignoring those measurements during which the tight-link becomes idle, it exploits the correlation that still exists between those dispersion measurements and the bursty cross-traffic under high variability conditions.

2   Neuro-fuzzy Cross-Traffic Estimator

Consider two probing packets of length L bits, sent T seconds apart over a FIFO link of capacity C bps, so that the queue does not empty between the departure of the first packet and the arrival of the second packet. The dispersion D will be L/C plus the time taken to transmit the cross-traffic that arrived during T. Consequently, we can estimate the average cross-traffic rate during that period of length T as $X = (D \cdot C - L)/T$. If there is no cross-traffic or the cross-traffic is small enough to be completely transmitted between probe packet arrivals, both probe packets will find an empty queue and the dispersion will be $D = T$. In any other case, when one or both probe packets find a non-empty queue but there are empty periods between the departure of the first packet and the arrival of the second packet, the dispersion is a random variable more or less correlated with the cross-traffic process, depending on the fraction of time the link was idle.

Figure 1 shows the results of a simple simulation experiment where we used the estimator above on the Bellcore traffic trace BCpAug89 [7] when it shares a T1 link with probe packets sent every second. The upper plot shows the true and estimated traffic during a small period, while the lower plot shows the corresponding buffer length. The estimator is exact when the link is 100% busy (900 to 940 seconds) and poor when the link is almost completely idle (990 to 1030 seconds). An intermediate performance corresponds to a not too busy link (1065 to 1085 seconds).

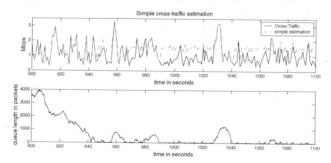

Fig. 1. Performance of the simple estimator

Let μ_D and σ_D^2 be the mean and the variance of the previous k dispersion measures. By observing Figure 1, we can establish plausible rules, such as "If μ_D is far from T, the simple estimation is exact", "If μ_D is close to T and σ_D^2 is small, the simple estimation is a poor one", and "If μ_D is close to T and σ_D^2 is large, the simple estimation is fair". This reasoning calls for a fuzzy approach to our estimation problem, for which we use a neuro-fuzzy system that allows us to construct an estimator suited to our particular training data (D_n, X_n), where D_n is the nth dispersion measurement and X_n is the corresponding cross-traffic rate. We decided to use D_n, D_{n-1}, and the sample mean and variance of the last 12 measurements as input variables, because they form a small set that have almost as much mutual information with X_n as the joint 12 dispersion measurements together. Next, traffic is centered and normalized with respect to the known capacity, C, while the dispersion measurements are centered and normalized with respect to T,

$$x_n = \frac{X_n - C}{C}, \qquad d_n = \frac{D_n - T}{T} \qquad (1)$$

So that the selected inputs to the traffic estimation system become

$$\theta_1(n) = d_n \qquad \theta_2(n) = d_{n-1} \qquad \theta_3(n) = \frac{1}{12}\sum_{k=0}^{11} d_{n-k} \qquad \theta_4(n) = \frac{1}{11}\sum_{k=0}^{11}\left(d_{n-k} - \theta_3(n)\right)^2 \qquad (2)$$

and the estimator $X_n = (D_n \cdot C - L)/T$ becomes the following simple estimator (SE)

$$\hat{x}_n = d_n - L/(C \cdot T) \qquad (3)$$

Fitting the histograms of each input variable conditioned on an "exact" or "poor" performance of the SE, we define fuzzy sets "Far from zero" and "Close to zero" through the following membership functions,

$$\mu_{CLOSE_i}(\theta_i) = \exp(-\lambda_i |\theta_i|), \quad \mu_{FAR_i}(\theta_i) = 1 - \mu_{CLOSE_i}(\theta_i) \quad \text{for } i=1,2,3 \text{ with } \lambda_1 = \lambda_2 \qquad (4)$$
$$\mu_{CLOSE_4}(\theta_4) = \exp(-\lambda_4 \theta_4), \qquad \mu_{FAR_4}(\theta_4) = 1 - \mu_{CLOSE_4}(\theta_4)$$

Now we use three rules for deciding whether SE is 'good', 'fair' or 'poor' and, for each case, we compute an affine transformation of θ_1 and θ_2. This way, we obtain only three non-linear parameters and nine linear ones, increasing regularity, simplifying the training process, and reducing the computational cost. We also use a simple simulation formula to estimate the queue length so that, if it is bigger than a given threshold (*thr*, an additional non-linear parameter), the SE is selected. The heuristically modified neuro-fuzzy estimator ($HNFE$) is shown in Figure 2.

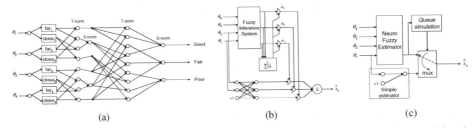

(a)   (b)   (c)

Fig. 2. The fuzzy inference system for classifying the quality of SE, (a), is used to combine the outputs of three appropriate affine estimators, (b). The queue heuristic is used to recover the simple estimator (c).

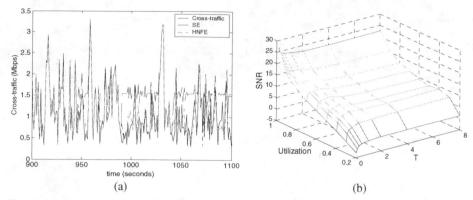

Fig. 3. Performance of the *HNFE* on the training data (a). SNR under different network conditions for a video cross-traffic trace (b).

In a preliminary step for training, we initialize λ_1, λ_3, and λ_4 by fitting the conditional histograms. The first step is the computation of the optimal linear parameters by a least square procedure. The second step is the computation of the optimal exponents through a quasi-Newton line search algorithm. These two steps are iterated until convergence. Finally, we look for the optimal *thr* through bracketing. The algorithm is so efficient that on the first evaluation of step 1, we already increased the signal-to-noise ratio (where the traffic trace is the signal and the estimation error is the noise) from 2 dB for *SE* to 9.3 dB for *HNFE*. After four iterations, we achieved 14.2 dB, and when we added the queue heuristic, we obtained a final performance of 14.9 dB on the training data. The whole procedure took a few seconds on a typical PC, for a one-hour traffic trace.

3 Numerical Results and Conclusions

Figure 3(a) reproduces Figure 1 with the new *HNFE* estimator. To check for generalization properties, we used the same *HNFE* estimator, without additional training, on a different traffic trace (a 768 kbps MPEG4 version of "Jurassic Park" [8]) under different link utilization factors, ρ, and different measurement periods, T. The estimation SNR is shown in Figure 3(b). The system exhibits a remarkable invariance with time scale and results very useful in a long range of traffic intensities, without any modification to the trained *HNFE*.

Concluding, in this paper we show that the high variability of modern networks traffic can be conveniently exploited for better instantaneous estimations within the range of time scales at which the high variability is exhibited. We devised a heuristically modified neuro-fuzzy cross-traffic estimator that is highly accurate, even for low long-run traffic intensities, as long as the variability allows for the presence of measurement periods with high link occupancy. This accuracy is achieved with very low computational complexity and at a minimal transmission overhead.

References

1. R.S. Prasad, M. Murray, C. Dovrolis, and K.C. Claffy, "Bandwidth Estimation: Metrics, Measurement Techniques, and Tools," IEEE Network Magazine, Vol. 17, No. 6, pp. 27--35, Nov/Dec. 2003.
2. N. Hu and P. Steenkiste, "Evaluation and Characterization of Available Bandwidth Probing Techniques," IEEE JSAC, Vol. 21, No. 6, pp. 879-894, Aug., 2003.
3. M. Jain and C. Dovrolis, "End-to-End Available Bandwidth Measure Methodology, Dynamics and Relation with TCP throughput," IEEE/ACM Transactions on Networking, Vol. 11, No. 4, August 2003, pp. 537-549.
4. V. Ribeiro, R. Riedi, R. Baraniuk, J. Navratil and L. Cottrell, "PathChirp: Efficient Available Bandwidth Estimation for Network Paths," Proceedings of Passive and Active Measurements (PAM) Workshop, La Jolla, CA, USA, Apr. 2003.
5. J. Strauss, D. Katabi, F. Kaashoek, and B. Prabhakar, "Spruce: A Lightweight End-to-End Tool for Measuring Available Bandwidth," Proceedings of Internet Measurement Conference (IMC) 2003, Miami, Florida, October 2003.
6. R. Ribeiro, M. Coates, R. Riedi, S. Sarvotham, B. Hendricks, and R. Baraniuk, "MultiFractal Cross-Traffic Estimation," Proceedings of ITC Specialist Seminar on IP Traffic Measurement, Monterey California, September 18-20 2000
7. Lawrence Berkeley National Laboratory. The Internet Traffic Archives, BC – Ethernet traces of LAN and WAN traffic, http://ita.ee.lbl.gov/html/contrib/BC.html
8. Video Traces Research Group, Arizona State University, http://trace.eas.asu.edu/TRACE/pics/FrameTrace/mp4/Verbose_Jurassic.dat.

Measurement Analysis of Mobile Data Networks[*]

Young J. Won[1], Byung-Chul Park[1], Seong-Cheol Hong[1], Kwang Bon Jung[1],
Hong-Taek Ju[2], and James W. Hong[1]

[1] Dept. of Computer Science and Engineering, POSTECH, Korea
{yjwon, fates, pluto80, jkbon, jwkong}@postech.ac.kr
[2] Dept. of Computer Engineering, Keimyung University, Korea
juht@kmu.ac.kr

Abstract. This paper analyzes the mobile data traffic traces of a CDMA
network and presents its unique characteristics compared to the wired Internet
traffic. The data set was passively collected from the backbone of a commercial
mobile service provider. Our study shows the highly uneven up/downlink traffic
utilization nature in mobile data networks along with small packet sizes, so
called the mice in the network. In addition, the relatively short session length
reflects the user behavior in mobile data network. We have also observed a
large amount of retransmissions on the backbone and analyzed the
consequences of such phenomenon as well as the possible causes.

Keywords: passive measurement, mobile data traffic analysis, CDMA network.

1 Introduction

The increase in availability of the mobile data services offers new means of
ubiquitous communication and entertainment. These services include multimedia
short message service (SMS), content downloading (e.g., ring tones, mp3, web-blog
update, e-book), on-line game, instant messaging, multimedia streaming, and many
more. Due to the high cost and unfavorable conditions in the wireless environment,
the traffic characteristics of the cellular networks for the data services are
distinguishable from those of the wired Internet traffic.

Recently a few studies have analyzed the traffic traces of the cellular networks –
CDMA 1xEVDO [3, 4, 5] and GPRS [2] - that are more suitable for carrying the data
traffic. It is important to note that these experiments were limited to a small scale
measurement study of the packet traces at the two end hosts. Their efforts were
concentrated on measuring the TCP performance metrics rather than understanding
the user behavior and the root cause of unusual traffic patterns. We differ from these
works in that our analysis investigates the unique or unusual traffic characteristics
reflecting the user and data service patterns.

[*] This research was supported by the MIC (Ministry of Information and Communication),
Korea, under the ITRC (Information Technology Research Center) support program
supervised by the IITA (Institute of Information Technology Assessment)" (IITA-2005-
C1090-0501-0018) by the Electrical and Computer Engineering Division at POSTECH under
the BK21 program of the Ministry of Education, Korea.

S. Uhlig, K. Papagiannaki, and O. Bonaventure (Eds.): PAM 2007, LNCS 4427, pp. 223–227, 2007.

2 Mobile Data Traffic Characteristics

The mobile data traffic trace we used in this paper is for a consecutive 12 hour period at the backbone of commercial CDMA network. More than 85% of the total packets are under the size of 100 bytes or less. This is a certainly distinguishable difference compared to the wired network. Such packets are less likely to transfer the actual data but to fulfill the signaling purpose of negotiating or maintaining the connection. The heavy tail of the flow length distribution theory might not be valid for the mobile data network.

The majority of the traffic is bound to the mobile devices from the content servers or WAP gateway - inbound. The ratios (%) between the inbound and outbound in packet counts and byte counts are 85:15 and 91:9, respectively. It follows a similar trail of the HTTP request-response behavior [1] which is somewhat an outdated traffic pattern for today's Internet.

Fig. 1 (a) and (b) indicate that over 90% of data traffic sessions are terminated within 10-20 seconds. In case of UDP sessions, the proportion of 0-10 seconds sessions are higher than those in TCP because of the unwanted DNS packets. Fig. 1 (c) illustrates the distribution of the flow lengths in the wired Internet. The available streaming and file downloading applications (e.g., P2P file sharing) in the wired Internet initiate longer data transmissions compared to the mobile data services where a light SMS transfer is yet the dominant application. The fundamental difference of the service applications and user behavior contribute to lowering the average session length in mobile data networks.

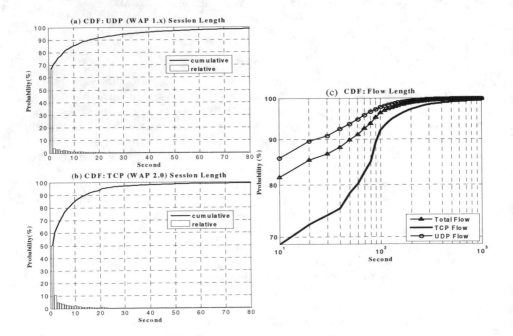

Fig. 1. CDF of mobile session lengths and flow lengths in the wired Internet

3 Retransmission Analysis

Fig. 2 illustrates the total and retransmission packet counts over the 12 hour period. Almost 80% of the total packets captured in the link are found to be retransmission packets. Sending unnecessary retransmission packets can cause a waste of network bandwidth and negatively influence the transparency of network usage billing.

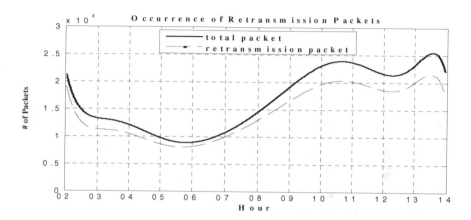

Fig. 2. Total packet vs. Retransmission packet counts

We have selected two sample TCP sessions in a close time interval. Session A experiences a large number of retransmission packets: 237 of the total 575 packets are retransmission packets between the time of 02:43 and 03:09. While session B consists of just one retransmission packet out of the total 237 packets between 02:41 and 03:01. These sessions are oriented between two separate mobile devices and one content server. Fig. 3 (a) and (b) illustrate the packets per second (PPS) measures over time for session A and B, respectively. The PPS for both sessions remains steady with value of 1; however, we observe a sudden peak at the almost end of session A – a few occurrences of 59 PPS. This phenomenon indicates the retransmission attempts from the content server, in which multiple copies of the same packet are sent continuously.

The total TCP sessions detected during the first half of our monitoring period was 235,568. About 77% of them encounter the packet retransmissions. Fig. 3 (c) illustrates the number of sessions sorted by their retransmission packet ratio to the total packet count of each session. The index on the x-axis represents the range of ratios (index 1 – 0 to 0.1, index 2 – 0.1 to 0.2, and likewise for the rest). Surprisingly, the sessions with retransmission ratio of more than 0.9 (index 10) is 38% of the total TCP sessions. In other words, 9 out of 10 packets are retransmission packets in these sessions. The majority of retransmissions in this experiment are less likely hardware duplicate because they have the pattern of occurring at the end of transmission.

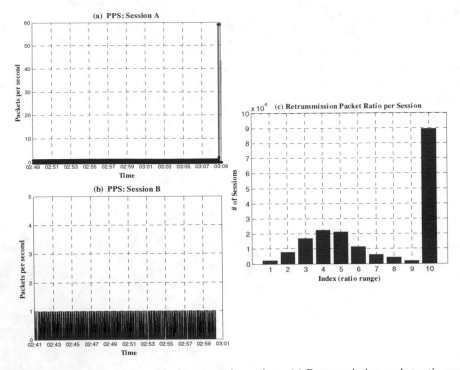

Fig. 3. (a), (b) Packets per second of two sample sessions; (c) Retransmission packet ratio per session

4 Concluding Remarks

This paper presented a study of the mobile data traffic characteristics by analyzing the data traffic trace from a commercial CDMA backbone network. A few unique traffic characteristics were observed: Uneven in/outbound traffic utilization at the mobile, a low average packet size, a short session length, and a high retransmission ratio. While comparing with the wired Internet traffic, we have found that the some of these characteristics followed an old trail of the wired Internet traffic. In addition, we have indicated unusually a large amount of retransmissions at the backbone and suggested possible answers to such phenomenon. Its consequences might raise a controversy of charging accuracy and performance degration in the mobile data services and also could help to troubleshoot the configuration problem at the service provider's network.

References

1. T. Kunz, T. Barry, X. Zhou, J.P. Black, and H.M. Mahoney. "WAP Traffic: Description and Comparison to WWW Traffic," ACM Workshop on Modeling, Analysis and Simulation of Wireless and Mobile Systems, August 2000.
2. Peter Benko, Gabor Malicsko, and Andras Veres. "A Large-scale, Passive Analysis of End-to-End TCP Performance over GPRS," IEEE INFOCOM, Hong Kong, March 2004.

3. Mark Claypool, Robert Kinicki, William Lee, Mingzhe Li, and Gregory Ratner. "Characterization by Measurement of a CDMA 1x EVDO Network," Proc. of the Wireless Internet Conference, Boston, MA, USA, Aug. 2006.
4. Youngseok Lee, "Measured TCP Performance in CDMA 1x EV-DO Network," PAM 2006, Adelaide, Australia, March 2006.
5. Wei Wei, Chun Zhang, Hui Zang, Jim Kurose, and Don Towsley. "Inference and Evaluation of Split-Connection Approaches in Cellular Data Networks," PAM 2006, Adelaide, Australia, March 2006.

Analysis of ICMP Quotations

David Malone[1] and Matthew Luckie[2]

[1] Hamilton Institute, NUI Maynooth
David.Malone@nuim.ie
[2] WAND Group, Computer Science Dept., University of Waikato
mjl@wand.net.nz

1 Introduction

RFC 792 requires most ICMP error messages to quote the IP header and the next eight bytes of the packet to which the ICMP error message applies. The quoted packet is used by the receiver to match the ICMP message to an appropriate process. An operating system may examine the quoted source and destination IP addresses, IP protocol, and source and destination port numbers to determine the socket or process corresponding to the ICMP message. In an idealised end-to-end Internet, the portion of the packet quoted should be the same as that which was sent, except for the IP TTL, DSCP, ECN bits, and checksum fields. In the modern Internet, this may not always be the case. This paper presents an analysis of ICMP quotations where the quote does not match the probe.

2 Methodology

2.1 Data Collection

Using `tcptraceroute`, the paths to 84393 web servers used in a previous study [1] were traced serially between the 6th and 12th of May 2005. All TCP SYN packets sent from the measurement source, as well as all ICMP time exceeded, unreachable, source quench, redirect, and parameter problem messages were recorded using `tcpdump`. 1190351 probes were sent, and 858090 ICMP replies were received and matched to a probe from 53768 unique IP addresses. 836456 ICMP responses were of type time exceeded, 21525 were of type unreachable, and 109 were of type source quench. A further 9 ICMP messages were unmatched.

By default, `tcptraceroute` generates a TCP SYN packet to port 80 and assigns the packet a unique IP-ID so that any subsequent response can be matched to its probe. The ECE and CWR TCP flags were set to test the reaction of middleboxes to these flags. The DSCP/ECN IP fields were set to 0x0f to identify systems that might modify these fields as part of the forwarding process. The DF bit was set to identify behaviour related to workarounds for broken path MTU discovery. Finally, each hop was probed once, to a maximum of 25 hops.

S. Uhlig, K. Papagiannaki, and O. Bonaventure (Eds.): PAM 2007, LNCS 4427, pp. 228–232, 2007.
© Springer-Verlag Berlin Heidelberg 2007

2.2   Quote Matching

As we may be matching an ICMP response to a packet that has been modified in flight, we are relatively liberal when matching an ICMP response to a probe. Based on the responses seen, the following heuristic was devised. A list of the 25 most recently sent probes is kept, as well as an array of the most recently sent probes for each IP-ID value. If an ICMP response can be matched by IP-ID or byte-swapped IP-ID to one of the 25 most recently sent probes, then it is deemed a match. Otherwise, it is not clear if any of these 25 probes match the ICMP response, because it is possible that either the IP-ID in the quoted packet was modified or the response was significantly delayed.

We score each of these 25 probes, as well as the last probes sent with a matching IP-ID or a byte-swapped IP-ID, and select the probe that meets the greatest number of the following criteria: matching destination IP address, matching TCP source port, matching TCP sequence number, no previous matching response, in last 1200 sent. Providing at least one of the IP-ID, destination IP address, TCP source port, or TCP sequence number matches, the probe with the largest number of matching criteria is the matching probe. To validate this technique, more unusual matches were inspected manually, and they appeared to be genuine.

2.3   Modification Classification

A modification may be classified one of three ways. First, if a modification is made to a single field and it appears unrelated to any other modification made, then it is noted as only affecting that field; the modification is further examined to determine if the field was set to zero, byte-swapped, incremented, or altered in some other way. Second, if a modification alters a set of fields in a related way, the modifications are summarised. For example, if an intermediate node inserts a TCP MSS option into a SYN packet as it is forwarded, then to do so correctly it will adjust the IP length field, TCP offset, TCP checksum, as well as include the option itself. Third, if a modification appears to be the result of accidentally overwriting a series of consecutive fields in the quoted packet, such that the integrity of the quoted packet is now compromised, the modification is classed as clobbering the fields.

2.4   Spacial Classification

Where possible, we infer if the modification is made in-flight while forwarding the probe, or is made during the quoting process and therefore localised. Quotes from a pair of adjacent hops are required to spacially classify a modification. A modification is associated with the first IP address of the pair. If a modification is observed at one hop but not the next, it is classed as a *quoter* modification. If the same modification is observed at adjacent hops, it is classed as an *in-flight* modification provided at least one of the corresponding IP or TCP checksums quoted is valid, indicating the change was intentional. This reduces the chance that adjacent quoter modifications are incorrectly classified as an in-flight

Table 1. Modifications made to IPv4 and TCP headers, by quoter IP address

Modification	In-flight	Quoter	Edge	Total Unique
IPTOS_MOD	1533 (2.9%)	146 (0.3%)	1674 (3.1%)	3030 (5.6%)
IPLEN_SWAP	0 (0.0%)	0 (0.0%)	1 (0.0%)	1 (0.0%)
IPLEN_MOD	0 (0.0%)	174 (0.3%)	322 (0.6%)	480 (0.9%)
IPID_SWAP	0 (0.0%)	29 (0.1%)	469 (0.9%)	494 (0.9%)
IPID_MOD	0 (0.0%)	1 (0.0%)	19 (0.0%)	20 (0.0%)
IPDF_MOD	4 (0.0%)	1 (0.0%)	30 (0.1%)	35 (0.1%)
IPOFF_SWAP	0 (0.0%)	32 (0.1%)	49 (0.1%)	80 (0.2%)
IPDST_MOD	29 (0.1%)	36 (0.1%)	1189 (2.2%)	1248 (2.3%)
TCPSRC_MOD	0 (0.0%)	3 (0.0%)	43 (0.1%)	46 (0.1%)
TCPDST_MOD	1 (0.0%)	2 (0.0%)	129 (0.2%)	132 (0.3%)
TCPSEQ_MOD	1 (0.0%)	0 (0.0%)	12 (0.0%)	13 (0.0%)
TCPACK_MOD	0 (0.0%)	0 (0.0%)	19 (0.0%)	19 (0.0%)
TCPMSS_ADD	4 (0.0%)	0 (0.0%)	19 (0.0%)	23 (0.0%)

modification, perhaps due to using the same router model with the same quirk at adjacent hops. Otherwise, if a modification is observed, but there is not a quote from an adjacent hop available for spacial classification, then it is classified as an *edge* modification. Finally, we stop processing a path when a loop is inferred.

3   Results

3.1   Observed Quote Lengths

Most quoters observed (87.60%) quote the first 28 bytes of the probe, which is the minimum amount permitted. 8.60% quote 40 bytes, corresponding to the size of the probe sent, and 2.14% quote 140 bytes, corresponding to ICMP quotations with MPLS extensions included. Therefore, at least 10.7% of quoters allow the complete IP and TCP headers of a probe to be compared with its quote.

3.2   Modifications to IPv4 Headers

The most frequent in-flight modification observed is to the DSCP/ECN byte: 2.9% of quoters were observed to modify this byte. 31 were inferred to use inconsistent values when overwriting the byte. This could be a measurement artifact due to attributing the change to the first IP address where it was observed, rather than an IP address of the previous hop where the change may have been made. 1073 quoters were observed to clear the DSCP, but leave the ECN bits intact, while 429 quoters were observed to clear the complete byte. 71 were observed to assign a DSCP of '001000', indicating some networks may prioritise HTTP traffic using the IP precedence bits. The second-most frequent in-flight modification observed was a modification of the destination IP address; of the 29 quoters that made modifications, 16 used RFC 1918 private addresses. Finally, 4 quoters were observed to clear the DF bit of an in-flight packet.

The quoter modifications observed on the IPv4 header indicate artifacts of processing the packet. The most frequent quoter modification observed is to the IP length field; of the 174 modifications, 160 were to change the field from having a value of 40 (0x28) to a value of 60 (0x3c), while the remaining 14 changed it to a value of 0x2814. Both modifications suggest the length of the IP header was added during processing; in the second case, the field was byte-swapped first.

3.3   Modifications to TCP Headers

Table 1 shows that most modifications to the TCP header were not identified as quoter or in-flight modifications; therefore, we examine all unique modifications. 46 quoters modified the TCP source port; 10 of these quoters used port 1, while the rest chose values that were only seen once. 132 quoters modified the TCP destination port. Some values were observed from multiple quoters and show signs of port redirection; for example, port 81 was seen from 11 quoters, and port 8080 from 13 quoters. Other port values chosen were seen once or twice.

23 quoters revealed that an MSS option was added in-line to the TCP header, probably to work around paths with broken path MTU discovery. 15 quoters revealed an MSS of 536 bytes had been set, 5 an MSS of 1460, and a series of quoters revealed cases of 1360, 1414, and 1436.

3.4   Quote Clobbering

Some probe modifications (not listed in Table 1) are due to inadvertent clobbering or modification of the quote in the quoting process. We group these modifications into four categories. 71 quoters (0.1%) quoted the first 28 bytes of a probe correctly, but then clobbered more than half the remaining bytes in the quote. 203 quoters (0.4%) quoted 60 bytes in the response; as the probe was only 40 bytes in size, the remaining 20 bytes were from a previous user of the memory, as described in US CERT note VU#471084. An additional 4 quoters over-quoted by 10 bytes, although at least the first six extra bytes were zero. Finally, 14 quoters over-quoted by 8 bytes, and set the last 16 bytes to zero.

3.5   Observed RTTs

Some probe packets required a long wait for an ICMP response. 56 required arrived over 10 seconds after the probe was sent, 34 more than 100 seconds, and 30 more than 300 seconds. The reason for these long round trips is not obvious; perhaps the probe is triggering link establishment and the probe is forwarded when the link is complete.

4   Conclusion

This paper presents a methodology for analysis of ICMP quotations, and uses it to analyse a dataset collected with tcptraceroute to a large number of web servers. Many in-flight changes are able to be attributed to known packet rewriting techniques. In the data collected for this paper, relatively few quoters are inferred to modify packets in-flight, or indeed to modify them during processing.

Reference

1. Medina, A., Allman, M., Floyd, S.: Measuring interactions between transport protocols and middleboxes. In: Internet Measurement Conference. (October 2004)

Correlation of Packet Losses with Some Traffic Characteristics

Denis Collange and Jean-Laurent Costeux

France Télécom R&D, 905 rue A. Einstein
06921 Sophia Antipolis Cedex, France
{denis.collange,jeanlaurent.costeux}@orange-ftgroup.com

Abstract. We show in this paper the correlation between the packet losses and some aspects of the behavior of ADSL users, more precisely some traffic characteristics of their TCP connections: duration, size and inter-arrival. We observe that when the loss rate increases, connections are shorter and their inter-arrival times increase. These shorter sizes are not only due to the interruption of connections but also to a certain form of self-censorship. We define different thresholds and a notion of sensitiveness, to describe the impact of the loss rates on these traffic characteristics.

Keywords: Passive performance measurements, anomaly detection, statistical correlation, ADSL.

1 Introduction

One of the major preoccupations of ISPs nowadays is to meet the performance requirements of their customers. They may then monitor their network and detect performance problems according to the perception of users. A first problem relies on defining end-to-end performance criteria and thresholds such that better measured performance ensures that customers are satisfied and that the network works properly. The detection method should be fast and take into account various sensitiveness. A possible method may be to assume that customers modify significantly their behavior when they get bothered by bad performance. They may interrupt active transfers, avoid or postpone some large transfers. If the loss rate is really high or the round-trip times very variant, the TCP protocol or the application may also abort the connections. Besides, some users may renew stopped transfers and then increase the arrival rate of transfers. They may also, on the contrary, wait for the end of an active transfer before launching new ones. A change in the users' behavior may thus be an objective criterion to detect a degradation of network performance.

Former traffic analysis about the impact of the performance on users' behavior considered the interruptions of connections ([1] and [2]). Other authors analyzed various models of this impact, but without referring to observations on an operational network ([3], [4] and [5]). Overall we found many papers proposing methods to improve the precision of loss measurements. But we did not find so many papers trying to correlate a change in this metrics to the reaction of end users.

S. Uhlig, K. Papagiannaki, and O. Bonaventure (Eds.): PAM 2007, LNCS 4427, pp. 233–236, 2007.
© Springer-Verlag Berlin Heidelberg 2007

In this study we analyze the influence of packet losses of TCP connections and the correlation between the performance and the users' behavior. We first present in section 2 the architecture of the network, the captures, and two simple rough methods to approximate end-to-end packet loss. We then show in section 3 the existence of a correlation between packet losses and some aspects of users' behavior, the flow durations and sizes and their arrival rate. We try to quantify this influence through various thresholds.

2   Measurements Settings

First of all, we present our collection infrastructure, based on a running ADSL access network. We then detail how we estimate packet loss from our traffic traces.

Our probe is located between a BAS (Broadband Access Server) and the first routers. It is a passive probe: capturing packet headers and generating absolutely no traffic on the observed network. This BAS multiplexes the traffic of many DSLAM, connecting more than 4000 clients. It is noteworthy that we capture all IP, TCP and UDP packet headers going through the BAS without any sampling or loss. The collected data thus represents a huge amount of traffic and may be representative of the traffic on the Internet, at least of the variety of the application used on the Internet. The analysis we present here refers to a complete week of measurements during February 2006.

From these traffic traces we compute and store in real time many indicators characterizing the traffic and the performance of TCP and UDP connections. We define and implement two methods to estimate the loss rate for each TCP connection from TCP/IP packet headers. The first method measures the proportion of what we call "desequencements" packets whose sequence number is lower than the sequence number of the last previous packet from the same TCP connection. The other method is based on retransmitted packets, i.e. when we see twice a sequence number. We call such events "retransmissions", even if some of them may be duplicated packet or if packets lost before the probe and retransmitted are not counted.

These estimation methods only apply on TCP connections. This is not restrictive to detect the state of the network since the TCP protocol represents more than 90% of the traffic volume. We intend to apply similar analysis on RTP/UDP traffic, on applications where a loss rate may be measured and that may be influenced by these packet losses. This is typically the case of voice or video over IP.

3   Connections Durations and Sizes

Fig. 1 presents the distribution of connections durations in seconds according to the desequencements rate given in the legend. We note similar results on retransmissions. We observe that most of the connections last more than 1000 seconds as long as the loss rate remains lower than 10^{-3}. For higher loss rates, the curves are gradually shifted towards the left. So the connections become shorter when the loss rate increases. We note that with loss rates larger than 10^{-2}, the curves remain close, between 30 and 100 seconds. We deduce from this figure a clear correlation between the distribution of the durations and the loss rate of a TCP connection. However, nothing on these curves indicates that high loss rates result in interrupting connections in progress. There may be actually several reasons to explain the reduction of the durations in case of bad performance, for example self-censorship of users. Other explanations are more inherent to the applications themselves. Some of them control their flow themselves. This is particularly the case of the peer-to-peer applications, which generate most of the traffic. They may thus obtain lower loss rates than, for example, short web requests often launched in parallel to download simultaneously the elements of a HTML page. We observe two thresholds on the loss rate, for retransmissions as well as for desequencements. These thresholds characterize three behaviors:

- Long connections with loss rates lower than 10^{-3}
- Short connections with loss rates greater than 10^{-2}
- Intermediate behavior with loss rates between 10^{-3} et 10^{-2}.

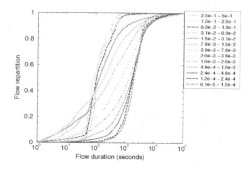
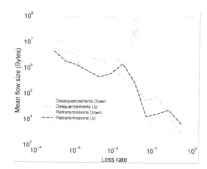

Fig. 1. distribution of flow durations for various desequencement rates

Fig. 2. Mean transfer size vs. loss rate

We represent on Fig. 2 the mean volume of connections in both directions (up and down) as functions of the desequencements and "retransmissions". As for duration, the size of connections is clearly related to their loss rates. We however remark that mean sizes are roughly divided by 1000 while durations are only divided by ten from the lowest to the highest loss rates. We observe a similar threshold of 4.10^{-3} (10^{-3} for durations) when the size of connections starts to decrease. The upper threshold when the size stops to decrease seems higher (10%). We note on retransmissions a slight increase of the mean size around 1% loss rate. As this peak does not appear with desequencements, it may be due to retransmitted packets. This may explain the lower decrease of durations because the retransmission delays are much larger than transmission delays, and they become preponderant in the connections' duration in case of high loss rates. It is thus more relevant to use the correlations between loss rate and transfer size rather than transfer durations, to estimate the influence of the end-to-end performance on users' behavior.

4 Conclusion

We highlighted and quantified the impact of the packet losses on many traffic characteristics of TCP connections. The packet losses appeared to be a good way to detect network performance problems and changes in users' behavior. Based on that fact, we noted thresholds on the loss rates, characterizing the impact on the distributions of the durations and sizes of the transfers. We then proposed in the complete study a method to characterize the sensitivity of traffic characteristics to losses. We applied it to compare the impact of losses on the transfer size and on the inter-arrivals of connections for various applications.

The detection of the alteration of users' behavior is not only the indication of bad network conditions, but also of dissatisfaction of users. Moreover, aborted transfers consume network resources unnecessarily, and thus disturb the successful transfers. The network operator must then correct the problem rapidly, in order to satisfy its customers as well as to optimize the use of the network.

References

1. M. Peuhkuri, "Internet Traffic Measurements - Aims, Methodology, and Discoveries", Helsinki University of Technology, Finland, 2002
2. D. Rossi, Ph.D. thesis "At the Network Edge: Sniff, Analyze, Act", Politecnico di Torino, Dipartimento di Elettronica, 2002
3. F. Guillemin, P. Robert, and B. Zwart, "Heavy tailed M/G/1-PS tails with impatience and admission control in packet networks", IEEE Infocom 03, San Francisco, CA, March 30 - April 3, 2003
4. J. Roberts and T. Bonald, "Congestion at flow level and the impact on user behaviour", Computer Networks 42 (2003) 521-536
5. S. Yang, G. Veciana, "Bandwidth sharing: the role of user impatience ", IEEE Globecom 01, pp. 2258-2262
6. V. Jacobson, C. Leres and S. McCanne, TCPdump, Lawrence Berkeley Laboratory, Berkeley, CA, June 1989
7. M. Allman, V Paxson and W. Stevens, "TCP Congestion Control", RFC 2581, April 1999
8. J. Padhye, V. Firoiu, D. Towsley and J. Kurose, "Modeling TCP throughput: A simple model and its empirical validation", ACM Sigcomm 1998, Vancouver, CA, September 1998, pp. 303-314
9. S. Jaiswal, G. Iannaccone, C. Diot, J. Kurose and Don Towsley, "Measurement and Classification of Out-of-Sequence Packets in a Tier-1 IP backbone", Infocom 2003
10. E. Brosh, G. Lubetzky-Sharon and Y. Shavitt, "Spatial-Temporal Analysis of passive TCP Measurements", IEEE Infocom 2005, March 2005, Miami, FL, USA
11. Y. Zhang, L. Breslau, V. Paxson and S. Shenker, "On the Characteristics and Origins of Internet Flow Rates", ACM Sigcomm 2002.

Investigating the Imprecision of IP Block-Based Geolocation

Bamba Gueye[1], Steve Uhlig[2], and Serge Fdida[1]

[1] Université Pierre et Marie Curie (Paris 6)
Laboratoire LiP6/CNRS - UMR 7606
{Bamba.Gueye,Serge.Fdida}@lip6.fr
[2] Delft University of Technology
Network Architectures and Services
S.P.W.G.Uhlig@ewi.tudelft.nl

Abstract. The lack of adoption of a DNS-based geographic location service as proposed in RFC 1876 has lead to the deployment of alternative ways to locate Internet hosts. The two main alternatives rely either on active probing of individual hosts or on doing exhaustive tabulation of IP address ranges and their corresponding locations. Using active measurements, we show that the geographic span of blocks of IP addresses make their location difficult to choose. Using the single location for a block of IP addresses as an estimation of the location of its IP addresses leads to significant localization errors, whatever the choice made for the location of the block. Even using as the location of a block the one that minimizes the global localization error for all its IP addresses leads to large errors. The notion of the geographic span of a block of IP addresses is fuzzy, and depends in practice very much on the uncertainty associated to the location estimates of its IP addresses.

Keywords: geolocation, active measurements, exhaustive tabulation.

1   Introduction

Location-aware applications have recently become more and more widespread [1, 2, 3]. One approach to locate Internet hosts is to push their location inside DNS records, as proposed in RFC 1876 [4]. Unfortunately, the adoption of this approach has been limited since it requires changes in the DNS records. There are also some geolocation services based on an exhaustive tabulation between IPs ranges and their corresponding locations. Examples of such services are *GeoURL* [1], the *Net World Map* project [2], and several commercial tools. Exhaustive tabulation is difficult to manage and to keep updated, and the accuracy of the locations is uncertain.

Padmanabhan et al. [5] developed three different techniques to map IPs to geographic locations and investigated the challenges in geolocation of Internet hosts. One of those techniques iteratively clusters IP addresses to map them to a single location. The authors of [5] observed that the accuracy of this method was related to the geographic spread of the hosts within these blocks of IP addresses.

In this paper, we quantify the extent to which locating all IP addresses within a block leads to an inaccurate geolocation of Internet hosts. We compare the location of blocks

S. Uhlig, K. Papagiannaki, and O. Bonaventure (Eds.): PAM 2007, LNCS 4427, pp. 237–240, 2007.

of IP addresses as given from two datasets [6, 7] with IP address location estimates based on active measurements. We show that the geographic span of the blocks of IP addresses, together with the intrinsic uncertainty of the exact location of individual IP addresses makes the choice of the location of a block difficult. The notion of the geographic span of a block of IP addresses is itself fuzzy, and depends in practice very much on the uncertainty associated to the location estimates of the IP addresses that belong to it. Even the *optimal location* (location that minimizes the global localization error of all IP addresses within a given block) of IP addresses blocks leads to significant differences between the estimated location of IP addresses and the one attributed to the entire block. Note that throughout this paper we refer to "block of IP addresses" as block and "IP addresses" as IPs.

The paper is organized as follows. Section 2 presents the datasets used to infer the location of target hosts based on active measurements. In Section 3, we investigate the inherent imprecision of estimating the location of individual IP addresses using a single location for their block. Finally, we conclude in Section 4.

2   Datasets

We consider two datasets containing IP or block of IP addresses to location entries in this paper. The first dataset contains $292, 362$ potential IPs of Web Clients that exchanged content over CoralCDN [6] and the second dataset is the database used by *GeoIP* [3]. For each IP address that composes the CoralCDN dataset, we seek its geographic location in the *GeoIP* [3] database. The GeoIP database provides also the block of IP addresses that this IP belongs to. Afterwards, we cross-check the location estimate obtained for each IP with its location estimate given by [7]. We found that $80, 449$ hosts provide different location estimates whereas $211, 913$ hosts have the same location estimate (at the city-level). Considering the IPs for which we found the same location estimate in the two databases, we apply the CBG technique [8] to find their geographic location estimate. For our measurements, we relied on 74 PlanetLab nodes spread all over the world as landmarks. During 3 weeks from 31 March until 19 April 2006, we conducted measurements to locate $25, 775$ IPs among the $211, 913$ whose location at the city-level agreed between the two databases. Among the set of IPs used, $7, 016$ have not been located by CBG. These hosts may be private, behind firewalls, or simply do not respond to *ping* probes. Thus we use for our study the remaining $18, 759$ IPs that were successfully located. The $18, 759$ successfully localized correspond to 876 blocks. The number of IP addresses probed within these 876 blocks varies between 3 and 197.

3   Limitations of Block-Level Geolocation

3.1   Geographic Span of IP Addresses Blocks

Estimating the actual geographic area spanned by a block of IPs is tricky. Geolocation of IP addresses based on active measurements and exhaustive tabulation both contain some uncertainty. In the case of active measurements like CBG, the geolocation is given in the form of an area where the host lies, the *Confidence Region* (CR) [8]. Since all IPs of a block are attributed to a single city-level location in the two used databases, it is impossible to estimate the span of blocks based on this information. Hence we have to rely on the estimates provided by CBG for each IP address. Note that we use as the location of an IP address the centroid of the CR computed by CBG [8].

For each block p, we compute the maximal distance between any two of its IPs $d_{max}(p)$ for which CBG gave us a location. We call $d_{max}(p)$ the *maximal span* of block p. Since $d_{max}(p)$ might be far larger than the typical distance between the locations of any two IP addresses within p, we also compute the median of the distance between any pair of IP addresses within p. We call this median distance the *median span* of block p.

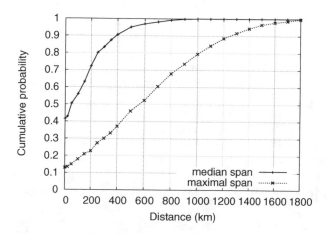

Fig. 1. Geographic span of blocks

Figure 1 provides the CDF of both the maximal and the median spans over the 876 blocks of IPs. More than 10% of the blocks have a maximal span of 0, i.e. all their IPs have exactly the same location. More than 40% of the blocks have a median span of 0 Km. Half of the IPs of these blocks are located at the same spot. Note that having a very small span for a block requires that the uncertainty of the geolocation of its IPs be very small, which typically happens when the localized host is close to one or several landmarks. About 50% of the blocks have a maximal span larger than 500 Km. Only 5% of the blocks have a median span larger than 500 Km.

3.2 Optimal Location of Blocks of IP Addresses

Assume that we want to have a location of a block that lies as close as possible to the locations of all IPs within this block. If we locate an IP at the centroid of the CR given by our active measurements, how large is the minimal geolocation error that we can expect when using as an estimate of the location of the IPs the location of the whole block? To answer this question, we compute the optimal location for each block of IP addresses, i.e. the location that minimizes the sum of the distances between the location of the block and the centroid of the CR of each of its IPs. If we were to do that, we would obtain approximation errors as the one shown on Fig. 2.

If we attribute to each block the optimal location, only a little bit more than 40% of the IPs would be located at most 200 km away from their estimated location based

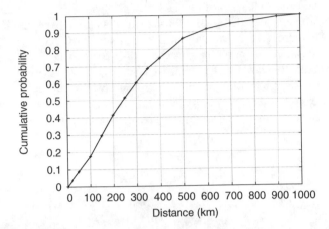

Fig. 2. Distribution of distances between optimally-located blocks and their IPs

on active measurements. More than 10% of the IPs would still suffer from a wrong localization more than 500 km away from their estimated location. Even though this is a little better than the localization we have in the database, this would still be far from satisfactory compared to the localization active measurements are able to provide.

4 Conclusion

We investigated the imprecision of relying on the location of blocks of IP addresses to locate Internet hosts. We showed that the geographic area spanned by blocks can be large, far larger than the typical distance between any two IPs within a block. We showed that even using the optimal location of a block leads to large geolocation errors. Our work indicates that it is necessary to assess the quality of geolocation information coming from exhaustive tabulation, because it contains an implicit imprecision.

References

1. *GeoURL*, http://www.geourl.org/.
2. *Net World Map*, http://www.networldmap.com/.
3. MaxMind LLC, *GeoIP*, http://www.maxmind.com/geoip/.
4. C. Davis, P. Vixie, T. Goodwin, and I. Dickinson, "A means for expressing location information in the domain name system," *Internet RFC 1876*, Jan. 1996.
5. V. N. Padmanabhan and L. Subramanian, "An investigation of geographic mapping techniques for Internet hosts," in *SIGCOMM*, San Diego, CA, USA, Aug. 2001.
6. M. J. Freedman, E. Freudenthal, and D. Mazires, "Democratizing content publication with coral," in *Proc. of USENIX NSDI*, San Francisco, California, March 2004.
7. M. Freedman, M. Vutukuru, N. Feamster, and H. Balakrishnan, "Geographic locality of IP prefixes," in *Proc. ACM/SIGCOMM IMC*, Berkeley, CA, USA, Oct. 2005.
8. B. Gueye, A. Ziviani, M. Crovella, and S. Fdida, "Constraint-based geolocation of internet hosts," *IEEE/ACM Transactions on Networking*, to appear.

A Live System for Wavelet Compression of High Speed Computer Network Measurements

Konstantinos Kyriakopoulos and David J. Parish

High Speed Networks, Electronic and Electrical Engineering,
Loughborough University, Loughborough, Leicestershire LE11 3TU, U.K.
{k.kyriakopoulos, d.j.parish}@lboro.ac.uk
http://www-staff.lboro.ac.uk/~elkk/

1 Introduction

Monitoring high-speed networks for a long period of time produces a high volume of data, making the storage of this information practically inefficient. To this end, there is a need to derive an efficient method of data analysis and reduction in order to archive and store the enormous amount of monitored traffic.

Satisfying this need is useful not only for administrators but also for researchers who run their experiments on the monitored network. The researchers would like to know how their experiments affect the network's behavior in terms of utilization, delay, packet loss, data rate etc.

In this paper a method of compressing computer network measurements while preserving the quality in interesting signal characteristics is presented. Eight different mother wavelets are compared against each other in order to examine which one offers the best results in terms of quality in the reconstructed signal. The proposed wavelet compression algorithm is compared against the lossless compression tool bzip2 in terms of compression ratio (C.R.). Finally, practical results are presented by compressing sampled traffic recorded from a live network.

2 Methodology

Wavelet analysis is not a compression tool but a transformation to a domain that provides a different view of the data that is more eligible to compression than the original data itself. This happens because small wavelet coefficients can be discarded without a significant loss in the quality of the signal. On the other hand, large coefficients represent important characteristics of the signal and they should be kept.

Gupta and Kaur [1] proposed an adaptive thresholding technique that is calculated from the value of the wavelet coefficients. This scheme is not based on signal denoising but rather tries to statistically identify significant coefficients.

Afterwards, normalization and run length encoding are applied. For the simulation experiments thirty delay and thirty data rate signals of 1024 points were used. The delay signals were measured over the test bed of High Speed Networks (HSN) research group. The data rate signals are from a real commercial network.

S. Uhlig, K. Papagiannaki, and O. Bonaventure (Eds.): PAM 2007, LNCS 4427, pp. 241–244, 2007.
© Springer-Verlag Berlin Heidelberg 2007

3 Wavelet Comparison

Eight wavelets were chosen and compared against each other in order to find out which one offers better reconstruction results. The following wavelets were compared: Haar, Meyer, Biorthogonal 3.9 and Daubechies D4, D6, D8, D10, D12. The index of Daubechies wavelets indicates the number of coefficients. The number of vanishing moments each Daubechies wavelet has is half of the number of coefficients.

Wavelets with many vanishing moments are described with many coefficients in the scaling and wavelet functions, thus increasing the computation overhead of the wavelet transform, the complexity of the algorithm and the output file size. Table 1 shows the average PSNR value after reconstruction at level 6 for thirty delay and data rate signals.

Table 1. Average PSNR for delay and data rate signals after reconstruction at level 6

Wavelet	Haar	D4	D6	D8	D10	D12	Meyer	Bio3.9
PSNR (db) Delay	39.60	38.25	37.65	37.47	37.05	36.97	37.08	37.35
PSNR (db) Data Rate	55.16	54.06	53.99	50.69	52.59	53.02	54.91	51.72

The Haar wavelet provides higher PSNR values for the reconstructed signals in both delay and data rate signals and has the following advantages: It is conceptually simple, fast, memory efficient and exactly reversible without producing edge effects.

4 Simulation Results

Fig. 1a shows a delay signal, before and after the compression. Because the two signals are very similar, the error between them is also provided for better judgment (lower line). The signal is decomposed at level 10 and the reconstruction quality is 37.85 dB while the C.R. is 13.7. PSNR values less than 35 dB loose some of the important signal characteristics while PSNR values less than 30 dB are not acceptable for such signals.

Fig. 1b shows a more interesting case of a data rate signal. This signal includes a spike, which is kept intact after the compression. A characteristic of the proposed algorithm is that it detects the spike as a more interesting feature than the rest of the signal. As a result, the algorithm's first priority becomes to preserve this characteristic and then comes the rest of the signal. The PSNR is 35 dB and the C.R.= 26.57.

Fig. 2 compares the C.R. results of the suggested wavelet technique against bzip2. It is interesting to examine the results from wavelet transform against a non-transform compression technique. In average, for delay signals (Fig. 2a) the suggested method (WT) achieves compression 6.5 times more than bzip2 with the best score being 11 times and the worst score 2.3 times. For data rate signals (Fig. 2b) the average compression is 4.7 times more than bzip2 with the best score being 12 times and the worst 4 times.

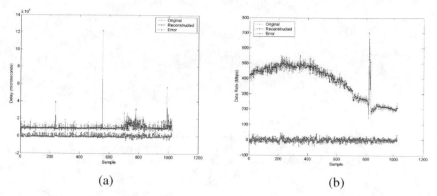

(a)   (b)

Fig. 1. (a) Delay signal 30 decomposed at level 10, PSNR= 37.85 dB. (b) Data rate signal 16 analyzed at level 5 with PSNR = 35.4 dB.

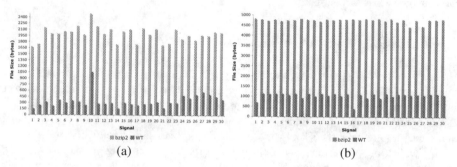

(a)   (b)

Fig. 2. Compression performance of the wavelet algorithm against bzip2 for delay (a) and data rate (b) signals. Each examined signal is located on the x-axis. The y-axis shows the file size in bytes.

5 Practical Results

The full algorithm is already implemented in CoMo. CoMo is a passive monitoring platform developed for the purpose of monitoring network links at high speeds and replying to real-time queries regarding network statistics. CoMo has various modules that each calculates one or more network measurements [2]. The proposed algorithm can be imbedded in the modules and compress these measurements. When CoMo receives a query, the information is first decompressed and then shown to the end user.

The experiment lasted for 8 days and CoMo was monitoring traffic recorded at HSN research group's live network. The overall achieved compression is 34.5 times. Fig. 3 presents a segment of 34 minutes from the 8 days duration experiment. This signal is characterized by discrete bursts of data rate. Some have amplitude of 70 kB/s while others are half that size or less. The reconstruction keeps intact the peaks and smoothes out the relatively small variation of the signal. PSNR for that segment is 55.9 dB.

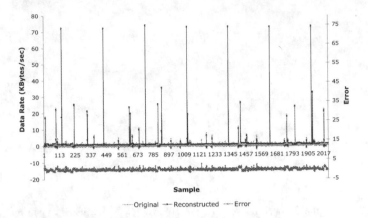

Fig. 3. Segment of 34 minutes of a data rate signal compressed live by CoMo. Error is given on the secondary y-axis on the right.

6 Conclusions – Future Work

This paper proposes the use of wavelet analysis techniques along with a wavelet coefficient thresholding method for compressing computer network measurements such as data rate and delay. Even though the compression is lossy, the important characteristics of the examined signal are preserved. In order to increase the compression, the detail characteristics are smoothed out by discarding the corresponding detail coefficients.

An evaluation of various wavelets with increasing vanishing moments was presented in order to determine which wavelet is more appropriate for performing the analysis. From simulation results, the Haar wavelet is found to be the best option as it offers the best results in terms of quality and compression ratio.

However, some improvements should be done in how the algorithm deals with the threshold in cases that spikes occur in an already bursty signal like in signal 16 (Fig. 1b). This would lead to more control over the quality of the reconstructed signal and more consistent PSNR values.

References

1. Savita Gupta and Lakhwinder Kaur, "Wavelet Based Image Compression using Daubechies Filters", In proc. 8th National conference on communications, I.I.T. Bombay, NCC-2002
2. Gianluca Iannaccone, Christophe Diot, Derek McAulley, Andrew Moore, Ian Pratt, Luigi Rizzo, " The CoMo White Paper", INTEL research technical report

Accurate Queue Length Estimation in Wireless Networks

Wenyu Jiang

Dolby Laboratories
100 Potrero Ave, San Francisco, CA 94103
wzj@dolby.com

Abstract. We describe a method of estimating the instantaneous queue length of a router or packet forwarding device in the last hop of a wireless network. Our method is more general and more accurate than QFind, and utilizes protocol knowledge of wireless networks such as 802.11 to achieve higher accuracy under both more general traffic conditions and less-than-ideal signal reception conditions during measurements.

1   Introduction

A router's queue length is an important indicator of network congestion, or lack thereof. A router's queue capacity, influences the quality of network services.

In [1], Claypool et al. describe QFind, a method of estimating the queue capacity of a last-hop router in an access network. It initiates a bulk download from a remote server to the measuring client, then it uses the product of delay increase in probe packets (D_q, the difference between maximum probe packet delay D_t and baseline probe packet delay D_l) and bulk download throughput T as an estimate of the queue capacity in bytes. Then the queue capacity in packets is $q_p = D_q \times T/s$, where s is the packet size of those traffic used to "overflow" the router's queue during a measurement. QFind has some limitations. First, it assumes that only a single application, initiated by the measuring node, is tieing up the router's queue. This is not always true, especially in cable modem type of access networks. Second, T is an average. This limits the accuracy of measurements, and does not allow estimation of the instantaneous queue length.

We present a method that solves these problems. In the remainder of this paper, we will describe and evaluate our method, and compare it with QFind.

2   Description of Queue Length Estimation Method

2.1   Estimation Procedure

Our method derives its estimate based on the definition of queue length. As illustrated in Figure 1(a), when packet P arrives at the wireless AP and enters the queue, there are L packets preceding it, i.e., the queue length is L. When packet P eventually makes its way to the head of this queue and exits the wireless

S. Uhlig, K. Papagiannaki, and O. Bonaventure (Eds.): PAM 2007, LNCS 4427, pp. 245–249, 2007.

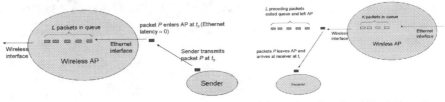

(a) Packet P enters AP at time t_s (b) Packet P exits AP at time t_r

Fig. 1. Status of wireless AP queue

AP, exactly L packets would have exited the AP on the same (wireless) network interface, as evidenced in Figure 1(b).

This is basically how our queue length estimation method works. All we need is a server that sends a probe packet P to the wireless measuring node, embedded with a sending timestamp that allows the measuring node to know when to start counting the packets that exited the AP, and stop counting at the time packet P is received, the resulting packet count will be exactly the instantaneous queue length when packet P arrives at the router. To measure the queue capacity, all one needs to do is initiate a bulk UDP flow that is guaranteed to saturate the link, and measure the resulting queue length. However, to ensure an accurate packet count, the following steps must be taken:

1. Filter the packets to only count packets emitted by the respective router/AP.
2. Deal with potential bit errors in observed packets, especially for important headers involved in the packet filter of step 1.
3. The clocks on both the sender and receiver (measuring node) must be synchronized to calculate a precise time window for counting packets.

2.2 Packet Filtering

For 802.11[2], all MAC packets or frames have a 2-bit "Direction" flag in their headers, consisting of a "FromDS" and a "ToDS" bit, where "DS" means an AP or router. The MAC addresses in a MAC frame, combined with this flag, reveal the sender and receiver. Below are the packet filters for counting packets emitted by an AP, or by a device other than the router/AP, respectively:

```
(1) FromDS==1 && ToDS==0 && Address2==MAC_AP_in_question
(2) FromDS==0 && ToDS==1 && Address1==MAC_AP_of_device_in_question
    && Address2==MAC_device_in_question
```

2.3 Dealing with Bit Errors

If the Direction flag is corrupted, we would have counted packets when they shouldn't be included and vice versa. Other important headers such as SEQCTL (sequence control) number, used to uniquely identify packets, are also vulnerable to bit errors when counting packets. In [3], we describe our zero-redundancy bit error correction method, by using implicit data redundancy in 802.11 MAC frames. For example, a MAC address with 1 bit error is easy to correct if no other valid MAC addresses exist in the network with a Hamming distance ≤ 1. This allows a more robust result in packet counting and similar procedures.

2.4 Queue Idleness Detection Method and Clock Synchronization

QFind uses the lowest one-way or round-trip delay to establish as a baseline delay for synchronization. However, QFind cannot verify if the queue is indeed idle when the lowest delay is measured. We have developed a method that detects whether a router/AP's queue is really idle, by reasoning with 802.11 CSMA/CA protocol interactions. In brief, our method says, if the channel has been idle for at least $DIFS + (CW_{min} + 1) \times timeslot_{802.11}$ prior to the *first* transmission of a certain packet, then the queue of the device that transmitted this packet must have been empty before this packet arrived in its queue.

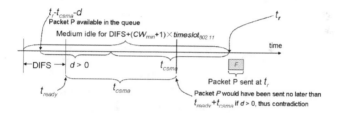

Fig. 2. Proof of AP/router queue idleness by contradiction

Figure 2 shows how to prove our method by contradiction. Basically, if packet P had experienced some queuing delay $d > 0$, packet P would have been received earlier than it is actually received. The details of the proof is omitted for brevity.

3 Performance Evaluation

We implemented our queue length estimation method for 802.11 by modifying the madwifi device driver on the Linux platform in an 802.11 testbed.

Our first test is to verify our method, and determine whether a router/AP's queue is byte-bounded or packet-bounded. We do this by sending a UDP cross traffic flow, with 3 different packet sizes, each lasting 5 seconds and enough to saturate the AP's queue. We tested it on a Netgear 802.11g AP. The result is in Figure 3(a) below, which clearly shows that this router's queue is packet-bounded, because when the byte size of cross traffic packets changes, only the queue length reported in number of packets remains nearly constant.

Next we used an AP with a lower transmission power to more easily test the effect of low SNR within a physically small testbed. We performed a test with 2 competing flows. A 30-second UDP flow and a 10-second TCP bulk download

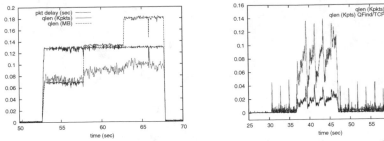

(a) Effect of cross traffic packet size (512, (b) Effect of multiple cross traffic flows, our
1024 and 1400 bytes, respectively)   method vs. QFind

Fig. 3. Effect of various factors on queue length estimation

flow. The average TCP throughput is then used in conjunction with the probe
packet delay to emulate QFind [1]. The results are shown in Figure 3(b). QFind's
result is clearly much lower than the actual queue length (since it does not
examine packets from other flows), which our method approximates well.

Thirdly, in Figure 4(a), the bottom/lowest curve shows what happens when
we only use MAC frames with a good CRC, the middle curve is when we we
simply ignore the CRC, but take a MAC frame *as is* (prone to miscounting
packets if some important MAC headers are corrupted), and the top curve using
our zero-redundancy error correction method. It is clear that our method gets
the best result, and closely follows the queue capacity threshold. Finally, in
Figure 4(b), we show that a SNR quality assessment algorithm (bottom curve)
we designed based on 802.11 SEQCTL number consecutiveness (higher value
indicates good SNR), can correlate well with actual SNR (top curve) and queue
length measurements (middle curve). This allows us to easily determine whether
our SNR is good enough to perform queue length or similar measurements that
require packet counting.

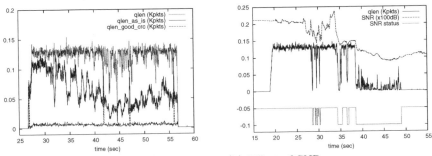

(a) Effect of 0-redundancy error correction (b) Effect of SNR

Fig. 4. Effect of SNR and error correction on queue length estimation

4   Conclusions

We present a method for accurately estimating queue lengths of devices with FIFO queues on a wireless or broadcast medium network. Through analysis and performance testing, we have shown that our method is more general and more accurate than QFind, and works much more reliably in less-than-ideal signal reception conditions. It can be used to enable cross-layer based Explicit Congestion Notification (ECN) mechanism in a last-hop network. It can be extended also to handle multiple FIFO queues, such as in the case of 802.11e, the QoS enhancement to 802.11, by adding the Traffic ID (TID) priority values as the matching criterion to our packet filter in Section 2.2.

References

1. Claypool, M., Kinicki, R., Li, M., Nichols, J., Wu, H.: Inferring Queue Sizes in Access Networks by Active Measurement. In: Passive & Active Measurement Workshop (PAM). (2004)
2. IEEE 802.11 Working Group: IEEE Std 802.11, Part 11: Wireless LAN Medium Access Control (MAC) and Physical Layer (PHY) Specifications. Technical report, IEEE (1999)
3. Jiang, W.: Bit Error Correction without Redundant Data: a MAC Layer Technique for 802.11 Networks. In: 2nd International Workshop on Wireless Network Measurement. (2006)

Measurement Informed Route Selection

Nick Duffield[1], Kartik Gopalan[2], Michael R. Hines[2]
Aman Shaikh[1], and Jacobus E. van der Merwe[1]

[1] AT&T Labs–Research
{duffield,ashaikh,kobus}@research.att.com
[2] Binghamton University
{kartik,mhines}@cs.binghamton.edu

1 Motivation

Popular Internet applications exhibit subtle dependencies on data path characteristics [3,5]. On the other hand, various studies [2,1] have shown that non-default Internet paths can have dramatically different quality characteristics compared to default paths. Unfortunately, the existing Internet routing infrastructure is not able to benefit from these observations. However, recently proposed routing architectures [4] open up the possibility to ease these constraints by allowing route selection to be more dynamic and to be informed by information outside the routing protocol. Motivated by these observations, in the work presented here, we explore the possibility of informing route selection based on measured path properties. Specifically, we observe that from their vantage points in the Internet topology, Tier-1 and other large ISPs typically have multiple possible routes that can be used to reach the majority of destination prefixes on the Internet. By monitoring routing information we track the availability of alternative paths available from a Tier-1 ISP. At the same time we perform detailed measurements of loss and delay to a large number of Internet destinations and characterize various properties of these alternate paths to determine: (i) Whether there are significant differences in the properties of these different paths that would warrant its consideration as part of the route selection process. (ii) The stability of these properties over various timescales which would impact how they can be utilized and dictate the requirements of a measurement infrastructure that can provide such information. We believe this is the first such study involving a single Tier-1 ISP.

2 Measurement Informed Route Selection

The essence of our approach is depicted in Figure 2. Consider source s_1 and destination d_1. s_1 connects to network $AS0$ at ingress router i_1. Network $AS0$ has two paths available to reach destination d_1, through egress router e_2 and network $AS1$ and through egress router e_3 and network $AS2$. In order to determine the "best" path, we assume the existence of a measurement infrastructure, such that performance measurements are available for the paths across $AS0$ between

S. Uhlig, K. Papagiannaki, and O. Bonaventure (Eds.): PAM 2007, LNCS 4427, pp. 250–254, 2007.
© Springer-Verlag Berlin Heidelberg 2007

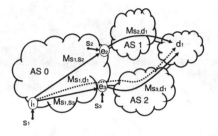

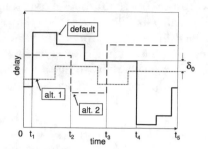

Fig. 1. Measurement Informed Route Selection: Composite measurements (s_1, s_2, d_1) and (s_1, s_3, d_1) and corresponding alternate paths

Fig. 2. Delay Measurement and Advantage: most recent measured delay on default and alternate paths. Advantage is $\geq \delta_0$ during interval $[t_1, t_4)$

the ingress router and the two egress routers, and from each of the egress routers to the destination. To simplify the notation, in what follows we will ignore the ingress and egress router notation and assume that a measurement source is co-resident with each of the routers in $AS0$. The set of measurements between the ingress router and the two egress routers is thus denoted by $M_{s1,s2}$ and $M_{s1,s3}$ respectively, and that between the egress routers and the destination by $M_{s2,d1}$ and $M_{s3,d1}$. Given this information, a network equipped with the appropriate routing infrastructure [4] can select the "best" route between s_1 and d_1 by appropriate combination of measured characteristics along the two available composite measurement paths (s_1, s_2, d_1) and (s_1, s_3, d_1). In this abstract we present initial results of a measurement study in which we evaluate the benefit of such Measurement Informed Route Selection as compared with the default BGP route selection as observed in a Tier-1 ISP.

3   Methodology

Composite Performance Metrics. Given loss and delay (λ_1, δ_1) on the path (s_1, s_2) and (λ_2, δ_2) on (s_2, d_1), the composite transmission rate for the composite path (s_1, s_2, d_1) is the product $1 - \lambda = (1 - \lambda_1)(1 - \lambda_2)$. The composite delay is the sum $\delta = \delta_1 + \delta_2$. The composite metric for loss $\lambda_{(s_1, s_2, d_1), t}$ at time t is the composite of the most recent measurements $\lambda_{(s_1, s_2), t}$ and $\lambda_{(s_2, d_1), t}$ on the internal and external segments, and similarly for delay δ.

Performance Advantage of Alternate Routes. A *route trajectory* of a source-destination (SD) pair (s, d) specifies for each time t a source $\sigma(t)$ which is s if the default path is used, and $s' \neq s$ if the alternate path (s, s', d) is to be used. The *loss advantage* of using a route trajectory σ is $\mathcal{L}(\sigma(t), t) = \lambda_{(s,d),t} - \lambda_{(s,\sigma(t),d),t}$, i.e., the difference between the most recent performance metric on the default path and the alternate path $(s, \sigma(t), d)$. (Here $\lambda_{(s,s,d),t}$ simply denotes $\lambda_{(s,d),t}$.) The delay advantage $\mathcal{D}(\sigma(t), t)$ is similarly defined.

Available Performance Advantage. The available performance advantage represents a baseline advantage that would be obtained by a routing policy that enabled instant selection of the best performing path whenever a new direct or composite measurement becomes available. Figure 2 illustrates delay measurements for a default path and two alternates. In each case, the curve shows the value of the most recent measurement on that path, and is hence a right continuous step function, the measurements occurring at the steps. In the interval $[t_1, t_2)$, alternate 1 has the best most recent performance measurement; in $[t_2, t_3)$ alternate 2 is best; in $[t_3, t_4)$ alternate 1 is best again; prior to t_1 and after t_4 the default route is best. Note the available delay advantage is positive, i.e., there is benefit in using a non-default path, only in $[t_1, t_4)$.

Temporal Performance Advantage Metric. We analyze the duration of *runs* of performance advantage, i.e., maximal time intervals in which the available performance advantage for a given SD pair exceeds a given level. Longer runs are more useful, since the payoff for switching routes is longer lived. If runs are shorter than the typical settle down time after route changes, the period of advantage would be over before its benefit could be utilized. Our delay run performance statistic is the *time fraction* $F_D(\tau, \delta)$: the fraction of the measurement interval that a given SD pair spent in a delay advantage run of duration greater than τ and whose performance advantage exceeds δ, for each $\tau, \delta > 0$. For example, with reference to Figure 2, measuring over $[0, t_5]$, then for any $\tau < \min\{t_4 - t_3, t_3 - t_2, t_2 - t_1\}$ and $\delta < \delta_0$, $F_D(\tau, \delta) = (t_4 - t_1)/t_5$. The corresponding loss statistic $F_L(\tau, \lambda)$ is defined similarly.

4   Evaluation

Performance Measurement. The data we used was obtained by performing active measurements continuously over a 12 day period in April 2006. The Internet paths to 738 unique destinations were probed from 15 probe locations distributed throughout the backbone of a large Tier-1 ISP. The probe destinations were randomly chosen from a significantly larger set of known DNS server addresses; hence measurement probes traveled via a diverse set of paths through the Internet. We also conducted measurements between all pairs of sources. For each SD pair, the loss measure λ was the proportion of 100 packets for which no response was received. The delay measure δ was the median reported round-trip time reported, including infinite values for lost packets. We favored the median over the mean, since it is more robust to outlying values.

Probe Frequency. Over 94% of SD pairs had median interprobe time < 20 minutes. The median time between receipt of probes from *any* source was < 2 minutes for over 97% of destinations. Thus composite performance data on alternate routes to a given destination is typically available within this timescale.

Median Performance. For each SD pair we calculated the median loss and delay across all probes. The median loss rate was 0 for about 92% of the pairs,

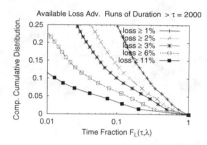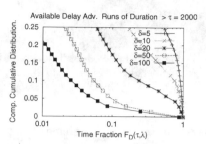

Fig. 3. Available Loss and Delay Advantage: CCDF of Time Fractions $F_L(\tau,\lambda)$, $F_D(\tau,\delta)$ over all SD pairs. Time fraction in which loss advantage exceeds threshold for $\tau = 2000$s. Left: loss. Right: delay.

with non-zero median loss rates distributed roughly uniformly between 0 and 1. About 95% of SD pairs had median delay less than 300ms. Between sources the median delay was roughly uniformly distributed between 0ms and 80ms.

Routing Data. We also determined BGP egress changes between every probe source and destination pair for the 12 day period, using BGP updates collected by a BGP Monitor from Route Reflectors in every PoP (Point of Presence) in the Tier-1 ISP. The updates allow us to determine the egress (the BGP next-hop) used by the route reflector for any destination. We mapped each probe source to the nearest PoP. We then used the updates collected from a Route Reflector in that PoP to determine egress changes from the source to every probe destination.

Number of Egress Points. Our framework assumes alternate routes exist. To test this, we computed the distribution of number of distinct egress points which advertise a given destination. The maximum number of egresses seen per destination was 10. For most destinations the number of egresses was constant for most of the duration trace (about 90% of the time). For example, about 64% of the destinations spend most of the time advertised by at least 7 egresses.

Performance Advantage Metric. We found the distribution of $F_L(\tau,\lambda)$, $F_D(\tau,\delta)$ over all SD pairs, for a range of parameters τ,λ,δ. Figure 3 illustrate these for runs lengths greater than $\tau = 2,000$s. The left plot shows the CCDF of $F_L(2000,\lambda)$ for loss advantages at least $(1\%,2\%,3\%,6\%,11\%)$. Consider 2% loss curve: about 15% of SD pairs spend at least 10% of the time in such runs lasting longer than 2,000s. The right plot shows $F_D(2000,\delta)$ for $\delta = (5,10,20,50,100)$ms. Consider the 20ms delay curve: about 17% of pairs spend about 10% of their time in such runs lasting longer than 2,000s.

Summary. Our initial results show the potential benefit of Measurement Informed Route Selection: (i) There is significant choice in terms of alternate paths to reach Internet destinations. (ii) Significant benefits in loss (at least 2%) and delay (at least 20ms) last for time periods that can be exploited by routing.

References

1. A. Akella, J. Pang, A. Shaikh, B. Maggs, and S. Seshan. A Comparison of Overlay Routing and Multihoming Route Control. In *Proc. ACM Sigcomm*, Sept. 2004.
2. D. G. Andersen, H. Balakrishnan, M. F. Kaashoek, and R. Morris. Resilient Overlay Networks. In *Proc. 18th ACM SOSP*, pages 131–145, Banff, Canada, Oct. 2001.
3. K.-T. Chen, P. Huang, G.-S. Wang, C.-Y. Huang, and C.-L. Lei. On the Sensitivity of Online Game Playing Time to Network QoS. IEEE Infocom, April 2006.
4. N. Feamster, H. Balakrishnan, J. Rexford, A. Shaikh, and J. van der Merwe. The Case for Separating Routing from Routers. FDNA Workshop, Aug 2004.
5. O. Tickoo, V. Subramanian, S. Kalyanaraman, and K. K. Ramakrishnan. LT-TCP: End-to-End Framework to Improve TCP Performance over Networks with Lossy Channels. IWQoS 2005, June 2005.

Fast, Accurate, and Lightweight Real-Time Traffic Identification Method Based on Flow Statistics

Masaki Tai, Shingo Ata, and Ikuo Oka

Graduate School of Engineering
Osaka City University
3–3–138, Sugimoto, Sumiyoshi-ku, Osaka 558–8585, Japan
{tai@n.,ata@,oka@}info.eng.osaka-cu.ac.jp

Abstract. Recently, identification of real-time traffic online is a key technology to achieve different service to real-time and bulk applications. Previously it is easy to identify real-time by checking the protocol/port number in IP header, however, it becomes more difficult due to the existence of streaming traffic over TCP connection, overlay networks such as P2P and VPN. In this paper, we propose a new identification method for real-time traffic based on not checking the protocol number, but analyzing the statistical characteristics of packet arrivals. Our approach is fast, accurate and lightweight compared to conventional techniques.

1   Introduction

Previously, it is easy to identify real-time traffic by checking the protocol number field in IP header because most of real-time applications use UDP packets. *Well-known port numbers* are also useful to classify the type of applications [1]. However, there is a possibility that the traffic classification by using such approaches will not be applicable in the future due to the existence of streaming traffic over TCP connection. Moreover, detection of real-time traffic over a kind of overlay network such as P2P or VPN is neither easy.

Recently, some approaches have tried to treat the problem of traffic identification. (e.g., signature-based classification [2,3] and machine learning techniques [4].) However, these approaches are not suitable for online identification because signature and ML based approaches require much computational overheads and data storages which are too huge to implement on network routers. The identification should be fast to follow up the speed of packet forwarding.

In this paper, we propose a new real-time traffic identification method which utilizes a remarkable difference on the statistics of packet inter-arrival gaps.

2   Monitoring Real-Time/Bulk Traffic and It Preliminary Analysis

We deploy a traffic monitor at the gateway of our laboratory. We mirrored traffic that passed the gateway of our laboratory, and capture all packets traversed

S. Uhlig, K. Papagiannaki, and O. Bonaventure (Eds.): PAM 2007, LNCS 4427, pp. 255–259, 2007.

Table 1. Summary of target traffic

Real-time		
Service	Source address	Protocol (Port number)
ISP	221.171.253.xxx	ms-streaming (1755)
News-MS	202.214.202.xxx	ms-streaming (1755)
News-Flash	202.214.162.xxx	macromedia-fcs (1935)
Music	167.167.9.xxx	ms-streaming (1755)
CM	58.157.27.xxx	macromedia-fcs (1935)
InternetTV	58.159.240.xxx	ms-streaming over rtsp (554)
Skype	61.196.29.xxx	skype (unspecified)

Bulk		
Service	Source address	Protocol (Port number)
Debian-FTP	203.178.137.xxx	ftp (21)
Vine	203.178.137.xxx	ftp (21)
Fedoracore	150.65.7.xxx	ftp (21)
Debian-HTTP	61.215.208.xxx	http (80)

between our laboratory and the Internet. We implement a capture program with `libpcap`. For each packet, source and destination IP addresses, source and destination port numbers, protocol number, of the packet and time stamp when captured are recorded.

After monitoring, we then classify captured packets into flows, and analyze the interval of packet arrival times in the flow. For the flow classification, we consider packets having the same five tuples as the same flow. We also set the timeout to detect the end of flow to be 60 seconds.

We analyze both real-time and bulk (i.e., data transfer) traffic by using TCP for classification without protocol and port numbers. In this paper, we analyze several streaming flows from famous sites in Japan as target real-time traffic. We also examine some file downloads from sites of Linux distributions as bulk transfer traffic. Sites and protocols are summarized in Table 1. Note that these protocols use TCP to deliver data traffic of streaming, i.e., we cannot identify real-time traffic by checking the port/protocol numbers. We also investigate some UDP applications like Skype VoIP traffic because we should be able to detect the UDP based streaming as well as TCP based one.

We especially focus on intervals of packet arrives in the flow. As a result of comparing time-variant intervals of packet arrivals, we can observe that intervals in bulk traffic are almost small, which indicates that arrival of packets in bulk traffic is bursty. On the other hand, for real-time traffic, the intervals of packet arrivals are observed by a kind of sawtooth wave, i.e., a set of packets is arriving with small gaps, and the interval between two sets is large.

3   Real-Time Traffic Identification Method

As a result of analysis in the previous section, we consider that comparing distribution of large gaps is an easy way to identify the real-time traffic. For this purpose, we classify traffics into bulk and real-time by following three steps. Prior to proceeding the identification, we classify traffic into flows by values of five tuples in the IP header.

1. Collect packet intervals having longer values by using LPF (Low Pass Filter).
2. Check whether the large gap is between the range of the interval which is used by most of real-time applications.
3. Check whether the intervals are appeared continuously.

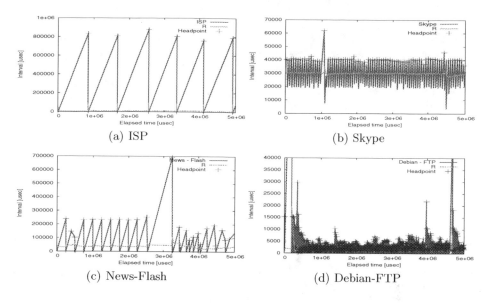

Fig. 1. Headpoint detection ($\beta = 1$)

3.1 Headpoint Detection by Using LPF

We define that the interval having the large value as *Headpoint* in this paper. To detect *headpoints*, we use a low pass filter (LPF) to eliminate small intervals. We apply the calculation of RTO (Retransmission TimeOut) of TCP as LPF. When the i-th packet of the flow arrives, The threshold T_i of LPF is given by $R_i \leftarrow \alpha R_{i-1} + (1-\alpha)M_i$ and $T_i \leftarrow \beta R_i$, where M_i is the measured value of interval between i-th and $(i-1)$-th packets. α and β are smoothing coefficient and coefficient of delay dispersion, respectively. In this equation, we use 0.1 for α. The interval M_i is detected as *headpoint* when $M_i > T_i$. Note here that the value of β strongly affects the accuracy of detection of *headpoints*. We therefore investigate the appropriate value of β. Fig. 1 compares *headpoints* detected when β is set to be 1.0 (i.e., the threshold is set to be the approximated average of intervals). As shown in this figure, we can observe that $\beta = 1$ achieves a good threshold to detect *headpoints*.

3.2 Range Check of Headpoints by BPF

When the j-th *headpoint* of the flow is detected, we check whether the interval of *headpoint* is generated by the real-time application or not. We use a band path filter (BPF) for checking. To obtain the appropriate range of BPF we investigate the distribution of packet intervals for each type of application.

We first calculate the interval time between two packets on a flow. Next, we sort all intervals in ascending order, and count the number of packets having the same interval. We finally obtain the cumulative probability of the distribution of intervals of packet arrivals. Especially, we focus on the region that the cumulative distribution is above 0.9. From this observation, intervals of real-time traffic are

also distributed from 20 msec to 1 sec, we can consider that generated by the
rate control of application. Our BPF marks *"Yes"* if the interval of *headpoint* is
in the range from 20 msec to 1 sec, otherwise the *headpoint* is marked by *"No"*.

3.3   Continuousness Check

After BPF marking, we finally check whether sequence of *headpoints* are continuously marked by *"Yes"*. If so, we identify the flow as real-time, otherwise we identify as bulk. However, this is quite rare that *headpoints* of bulk traffic are marked
"Yes" continuously due to the burstiness of the bulk traffic. Therefore we introduce
a counter which represents the number of *headpoints* continuously marked as *"Yes"*.
We use 5 (five) *headpoints* as the threshold of continuousness. That is, five continuous *headpoints* are within the range of BPF, the flow is identified as real-time.

3.4   Classification Algorithm

From previous subsections, we finally define the algorithm of real-time traffic
identification. First, arrived packets are classified into flows by checking five-
tuples of IP header. For each flow f, the algorithm stores four variables. Note
here that all of variables are initialized by zero.

- Previous value of packet interval M_f^p
- Threshold of LPF R_f
- Previous value of *headpoint* H_f^p
- Counter for continuousness C_f

When the i-th packet of the flow f arrives, the router perform following procedures.

1. Obtain the interval $M_{i,f}$
2. $R_f \leftarrow \alpha R_f + (1 - \alpha)M_i$
3. LPF check: if M_i is larger than R_f, $M_{f,i}$ is *headpoint* $H_{f,j}$ of flow f. Otherwise jump to Step. 7.
4. BPF check: if $H_{f,j}$ is between 20 msec to 1 sec, increment C_f by one. Otherwise let $C_f \leftarrow 0$, and jump to Step. 6.
5. Continuousness check: if $C_f = 5$, flow f is identified as real-time.
6. $H_f^p \leftarrow H_{f,j}$ and increment j by one
7. $M_f^p \leftarrow M_{f,i}$ and increment i by one

Mainly, our procedure consists of one calculation, three comparisons, and three
assignments. We consider that it is very small computational overhead which
enables to update online. Moreover, in the worst case, the detection time is 5 sec.
We consider that this delay is relatively small against the delay of buffering at
the beginning of playout in streaming applications.

4   Identification Results

To verify the accuracy of identification on our proposed method, we classify a publicly available traced data provided by CAIDA, which is captured data collected

at OC192c backbone link between the Indianapolis router node and Kansas City on June 1st, 2004. We can also observed from Table 2 that our method can identify streaming applications with high accuracy ($> 80\%$). Additionally, we classify locally traced data, which a different dataset from the one we used to analyze. During capture of the traffic, we played a number of real-time streaming traffic over HTTP and Skype VoIP traffic. As a result of identification on the data traced our laboratory, our proposed method can identify both real-time over HTTP and Skype VoIP traffic by checking the payload. For this reason, our method can also detect some streaming services over HTTP which are included in 6,534 flows.

Table 2. Identification results (CAIDA trace)

Category	Application/ protocol	# of flows identified as real-time		Detection rate
		Manually classified	Proposed	
Total	All	416933	26288	6.3%
Streaming	WindowsMedia	79	70	88.6%
Streaming	MacromediaFlash	36	30	83.3%
Streaming	Real, QuickTime	15	12	80.0%
Streaming	rtsp	419	389	92.8%
Web	http	36974	6534	17.6%
Data	ftp	7324	848	11.6%
Mail	smtp, pop, imap	6777	818	12.1%
Network management	dns, netbios, smb, snmp, ntp	82685	4129	5.0%

5   Conclusion

In this paper, we have shown that the arrival interval of the packet of the flow characteristic is obviously different between real-time traffic and bulk traffic. Next, we have proposed the algorithm to classify a real-time traffic and a bulk traffic automatically by using result of analysis. Experimental results have shown that our method can identify real-time traffic with high accuracy.

References

1. Andrew W. Moore and Konstantina Papagiannaki, "Toward the Accurate Identification of Network Applications," in *Proceedings of PAM Passive and Active Network Measurement*, pp. 41–54, March 2005.
2. Subhabrata Sen, Oliver Spatscheck, and Dongmei Wang, "Accurate, Scalable In-Network Identification of P2P Traffic Using Application Signatures," in *Proceedings of the 13th international conference on World Wide Web*, pp. 512–521, May 2004.
3. Thomas Karagiannis, Konstantina Papagiannaki, and Michalis Faloutsos, "BLINC: Multilevel Traffic Classification in the Dark," in *Proceedings of SIGCOMM Special Interest Group on Data Communication*, pp. 229–240, August 2005.
4. Andrew W. Moore and Denis Zuev, "Internet Traffic Classification Using Bayesian Analysis Techniques," in *Proceedings of the 2005 ACM SIGMETRICS international conference on Measurement and modeling of computer systems*, pp. 50–60, June 2005.

Impact of Alias Resolution on *Traceroute*-Based Sample Network Topologies

Mehmet Hadi Gunes, Nicolas Sanchis Nielsen, and Kamil Sarac

University of Texas at Dallas, Richardson, TX 75083
{mgunes,nicolas,ksarac}@utdallas.edu

Abstract. Most Internet measurement studies utilize traceroute-collected path traces to build Internet maps. In this paper, we measure the impact of alias resolution problem on Internet topology measurement studies. Our analysis shows that the alias resolution process has a significant effect on the observed characteristics of the topology maps.

1 Introduction

Internet measurement studies require the availability of representative Internet maps. Most measurement studies utilize a well-known Internet debugging tool called *traceroute* to collect a router-level topology map from the Internet. After collecting the path traces, the information needs to be processed to build the corresponding network topology. This step requires identification of the IP addresses belonging to the same router, a task often referred to as *IP alias resolution*. Since routers have multiple interfaces with different IP addresses, it is likely that a router may appear on multiple paths with different interface IP addresses. The goal of IP alias resolution is to combine the IP addresses that belong to the same router into a single node in the resulting topology map.

Several tools have been proposed to resolve IP aliases in traceroute-based topology construction studies. [1,2,3,4]. The current best practice is to use the existing tools to resolve IP aliases to build a topology map. However, there is no practical way to measure the success rate of the alias resolution process. That is, it is extremely difficult to collect the underlying topology information for verification purposes. In fact, the lack of the underlying topology information is the main reason that leads the researchers to conduct topology collection studies.

At this point, one issue is to understand the impact of the alias resolution process on the observed topological characteristics of the collected topology maps. If the impact is acceptably small, then we could have confidence on the conclusions of the measurement study even without an alias resolution process. On the other hand, if the impact is high, the conclusions of the study may significantly depend on the accuracy of the alias resolution process. Although several recent studies pointed out the impact of inaccurate alias resolution in certain measurement study results, to the best of our knowledge, there is no systematic study that quantifies the impact of inaccurate alias resolution on traceroute-based Internet measurement studies.

S. Uhlig, K. Papagiannaki, and O. Bonaventure (Eds.): PAM 2007, LNCS 4427, pp. 260–264, 2007.

In this paper, we present an experimental study to quantify the impact of alias resolution on Internet topology measurement studies. First we generate several synthetic network graphs using Barabasi-Albert (BA), Waxman (WA) and Transit-Stub (TS) network models. Then, we emulate traceroute functionality by collecting a number of path traces from the network graphs. During the sample topology construction, we use different success rates for the alias resolution process to obtain different sample topologies. Here, 0% indicates that alias resolution fails for all nodes in the network and 100% indicates that it succeeds for all nodes. We then study various topological characteristics of these sample topologies to quantify the impact of alias resolution on the observed results. We consider over 20 different graph characteristics including topology size, node degree, degree distribution, joint degree distribution, characteristic path length, betweenness, and clustering related characteristics. Due to size limitations, we present only a subset of the results to summarize our findings.

Our main conclusion in this study is that the accuracy of the alias resolution process has a significant impact on almost all topological characteristics that we consider in this study. Therefore, Internet measurement studies should employ all the means possible to increase the accuracy/completeness of the alias resolution process. Even in this cases, our confidence in the results of such measurement studies will be limited by the lack of a mechanism to verify the accuracy/completeness of the alias resolution process.

2   Impact of Alias Resolution on Degree Characteristics

In this section, we study changes in node degrees with improving alias resolution. In our experiments we observe that the accuracy of the alias resolution process has an important impact on the node degree-related characteristics of the sample topologies. Although one may intuitively expect an improvement on the accuracy of the degree-related characteristics with an increasing success rate of the alias resolution process, we may not necessarily observe such a trend all the time. Fig. 1 presents an example scenario where 'no-alias resolution' case (Fig. 1-b) results in a better approximation to (1) the degree of node a and (2) the average and the maximum degrees of the original subgraph (Fig. 1-a) compared to the 'partial alias resolution' case (Fig. 1-c) when we resolve aliases only for a.

We study sample topologies to observe the changes in node degrees as the success of the alias resolution process increases. This helps us gain more insight into the impact of the alias resolution process on the node degree characteristics.

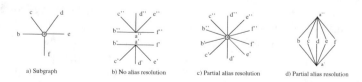

a) Subgraph   b) No alias resolution   c) Partial alias resolution   d) Partial alias resolution

Fig. 1. Effect of partial alias resolution on a subgraph

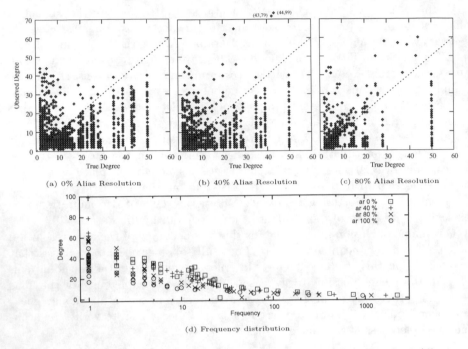

(a) 0% Alias Resolution   (b) 40% Alias Resolution   (c) 80% Alias Resolution

(d) Frequency distribution

Fig. 2. Degree comparison for (100,100)-sample topologies from Barabasi-Albert

Fig. 2-a,-b,-c show the changes in node degrees for (100,100)-sample topology of BA graph for 0%, 40%, and 80% alias resolution success rates. In these figures, 'Observed Degree' indicates the degrees of the nodes in the sample topology with imperfect alias resolution and 'True Degree' indicates the degrees in the sample topology with perfect alias resolution. Each point in these figures may correspond to one or more nodes in the sample topology with the same 'Observed' and the same 'True' degrees. The number of nodes corresponding to each point is presented in the frequency distribution graph in Fig. 2-d. As an example, the '+' tick at location (99,1) in Fig. 2-d indicates that there exists only one node with an 'Observed Degree' of 99 under 40% alias resolution success rate and this node is presented in Fig. 2-b with the point marked at (44,99) label at the top of the figure. The label indicates that the 'Real Degree' of this node is 44.

We now present several observations about the results presented in these figures. The points above the x=y line in Fig. 2-a,-b,-c correspond to overestimation of node degrees and the points below the x=y line correspond to underestimations of node degrees in the sample topologies. In general, overestimation is caused by alias resolution problems at the neighboring nodes of a given node. Fig. 1-c presents an example for this case. Similarly, underestimation is caused by alias resolution problems at the node itself. In addition, the comparison of Fig. 2-a,-b,-c show that the *observed* maximum degree of the graph first increases from 44 in Fig. 2-a to 99 in Fig. 2-b. It then goes down to 60 in Fig. 2-c (and down to 50 with 100% alias resolution success rate). Another observation from the figure is that alias resolution problems at a node may introduce a significantly large number of artificial nodes in the resulting sample topologies. As an example, according to Fig. 2-d, there is only one node with a *true degree* of 50 in the *real* sample graph (i.e., refer to (50,1) point in Fig. 2-d). On the other hand, Fig. 2-a,-b,-c show a large number of nodes with *observed degrees* less than 50 that correspond to a node with a degree of 50. Finally, we observe that as the alias resolution success rate increases some of the underestimation cases change to overestimation (compare Fig. 2-a vs. Fig. 2-b for x=43 and x=44 and Fig. 2-b vs. Fig. 2-c for x=35, x=37, and x=39). This indicates that although the alias resolution problems of the corresponding nodes are fixed, there exists some neighbors of these nodes with alias resolution problems causing overestimation.

3 Impact of Alias Resolution on Graph Characteristics

In this section we summarize the impact of poor alias resolution on topologies.

Topology Size: According to the experiment results, the success of alias resolution has a big impact on the topology size. Number of nodes and edges reduces 57% and 62%, respectively, on average for sample topologies as alias resolution improves from 0% to 100%. Besides, the impact of imperfect alias resolution increases as the size of the sample topology increases.

Degree Distribution: Degree distribution has been used to characterize network topologies and several topology generators use this characteristic to generate synthetic topologies. In our experiments, we observe that degree distribution changes with the changing success rate of the alias resolution process, but different effects are observed with different samples. For the power-law based graph samples, i.e., BA-based samples, imperfect alias resolution distorts the power-law characteristic of the distributions. For TS- and WA-based samples, the alias resolution process has different types of impacts especially at low degree or high degree ranges, respectively.

Characteristic Path Length: Characteristic path length (CPL) measures the average of the shortest path lengths between all node pairs in a network. In all of the sample topologies, CPL values reduce with the increasing alias resolution success rate. The average reduction for BA, and WA-based sample topologies is about 30%. For TS-based samples, we do not observe much changes. This is possibly due to the fact that the TS graph is a hierarchical graph and the shortest path lengths are not affected by the alias resolution process.

Betweenness: Betweenness is a mesure of centrality. It reports the total number of shortest paths that pass through node v. Usually betweenness is normalized with the maximum possible value, i.e, $n(n-1)$ where n is the number of nodes. We analyze betweenness distribution and observe considerable changes with the increasing alias resolution success rate. The average betweenness reduces with an improvement in alias resolution success rate. On the other hand, the normalized betweenness increases as the alias resolution success rate increases.

Clustering: Clustering characterizes the density of the connections in the neighborhood of a node. We analyze clustering distribution with respect to node degree and observe an increase with increasing alias resolution success rate. Clustering coefficient, a summary metric of clustering, is the ratio of the number of triangles to the number of triplets. In experiments, all samples yield a clustering coefficient of 0 with 0% alias resolution success rate and always increases with the increasing alias resolution success rate except for a single case.

References

1. Govindan, R., Tangmunarunkit, H.: Heuristics for Internet map discovery. In proc. of IEEE INFOCOM. (2000)
2. Gunes, M., Sarac, K.: Analytical IP alias resolution. In proc. of IEEE ICC. (2006)
3. Spring, N., Dontcheva, D., Rodrig, M., and Wetherall, D.: How to Resolve IP Aliases. University of Washington, Technical Report. (2004)
4. Spring, N., Mahajan, R., Wetherall, D., Anderson, T.: Measuring ISP topologies using rocketfuel. IEEE/ACM Transactions on Networking. **12** (2004) 2–16

Bridging the Gap Between PAMs and Overlay Networks: A Framework-Oriented Approach

Kenji Masui and Youki Kadobayashi

Nara Institute of Science and Technology
8916-5 Takayama, Ikoma, Nara 630-0192, Japan
{kenji-ma,youki-k}@is.naist.jp

1 Introduction

Besides the classic measurement methodologies such as *ping* for measuring RTT and *traceroute* for discovering IP topology, there also exists a new trend in measurement methodology, cooperative measurement [1,2]. In cooperative measurement, a measurement node sometimes communicates with other measurement nodes, shares collected data, and estimates the network characteristics of some parts of network elements without actual measurement. Cooperative measurement is considered appropriate especially for large-scale measurement on overlay networks because network characteristics can be helpful for increasing the autonomy of overlay networks and such measurement methodologies have a potential for the reasonable estimation of network characteristics against a number of elements within the limited measurement capacity of each node.

While we have a number of sophisticated passive and active measurement methodologies (PAMs) that provide informative network characteristics to overlay network applications, we are yet to witness the widespread adoption of such methodologies in these applications. One reason for such a situation is that the collection of network characteristics is often a burden on application developers, especially in the case of using sophisticated and complicated cooperative methodologies. Measurement is not developers' objective but just a means for refining their applications, and the extra load of implementing measurement methodologies takes precious time and prevents concentration on their original objectives. Although network characteristics are indispensable for sustaining overlay networks and we have various methodologies for the collection, this fact doesn't appeal to the developers — a gap exists between measurement and overlay.

In this paper, we present a general platform for large-scale distributed measurement named N-TAP, which provides APIs for obtaining network characteristics for overlay network applications. N-TAP is an independent service from the viewpoint of applications, and its APIs let application developers handle network characteristics easily. N-TAP is also a platform in which various measurement methodologies can be implemented. A measurement node of N-TAP can utilize its shared database and communication channels among measurement nodes for cooperation, and these features are achieved on the measurement overlay network of N-TAP. We believe the gap can be bridged by developing a software framework for both PAMs and overlay networks.

S. Uhlig, K. Papagiannaki, and O. Bonaventure (Eds.): PAM 2007, LNCS 4427, pp. 265–268, 2007.

2   N-TAP: A Platform for Large-Scale Distributed Measurement

N-TAP is designed with three concepts to be a solution to the gap problem described in Section 1. The first concept is "independent service." This concept means that N-TAP should abstract common measurement procedures to one independent service so that any kind of applications can utilize the measurement procedures easily. At the same time, an interface for the interaction between applications and N-TAP is required to be an independent service. The second one is "cooperative measurement," which accelerates the deployment of sophisticated methodologies by providing fundamental features for such methodologies. As the common features, N-TAP should prepare the mechanisms for communication channels among measurement nodes and shared databases of collected network characteristics. "Decision making" is the last one of the concepts. Since the requirement for network characteristics depends on each application, N-TAP should interpret such requirements carefully and make a decision on the action of collecting requested data. As far as N-TAP can fulfill application's requests, it should also consider the tradeoff that measurement methodologies contain among some various indices like measurement overhead, accuracy, scalability, timeliness and so on.

Based on these concepts, we designed the architecture of N-TAP and implemented its prototype [3]. The overall system of N-TAP consists of N-TAP agents, which are the daemon programs running on end nodes. An agent prepares an XML-RPC interface as an independent service in order to accept the requests from applications and provide the requested network characteristics. To implement cooperative measurement methodologies on N-TAP, the agents construct a Chord-based [4] measurement overlay network called N-TAP network, and provides the mechanisms to share collected data and communicate among the agents. For reducing measurement overhead and improving response time, the agent does "local-first, remote-last" decision making. With this rule, the agent utilizes past collected data as a response instead of performing actual measurement if they meet the requirements of network characteristics from applications.

Here, we describe one simple scenario of retrieving network characteristics from N-TAP. Suppose that there are three N-TAP nodes: nodes A, B and C. The application running on node C wants to know the RTT between nodes A and B that is collected within 60 sec., so it requests to the local N-TAP agent on node C by calling the method `ntapd.getNetworkCharacteristics.roundTripTime.IPv4`. Based on the process of decision making, the agent on node C first searches the data that meets the request in its local database. In this case, we assume node C cannot find such data in the local database. Since node C cannot measure the RTT between nodes A and B by *ping* of course, node C decides to search the data in a shared database. Unfortunately such data is not in the shared database, so node C confirms the existence of N-TAP node A by the list of N-TAP nodes in the shared database, and forwards the request to node A to measure it. Node A performs RTT measurement between nodes A and B, then replies to node

C with the result of the measurement. Finally, the agent on node C gives the requested data to the application with its response message.

Some systems such as iPlane [5] and S^3 [6] have almost same objectives with N-TAP from the viewpoint of the provision of network characteristics to applications. On the architectural aspect, N-TAP differs from these systems especially in the manner of data storage. Each N-TAP agent stores collected data in both local database and the shared database, and according to the "local-first and remote-last" rule, the agents prioritize the search in their local databases and the measurement on a local node for reducing measurement cost and improving response time. Another unique point is that N-TAP provides an programmable environment for cooperative measurement, which can be helpful for the evaluation of measurement methodologies.

3   Discussion

In this section, we discuss some topics that were revealed in the process of implementing N-TAP.

First topic is the merits and demerits of a framework-oriented approach. By using N-TAP as an independent service, we confirmed that applications can obtain network characteristics as easily as retrieving content from a web server. Moreover, thanks to the features for cooperative measurement, we can now obtain network characteristics that cannot be obtained by a solo node, such as a bidirectional IP topology between two nodes. On the other hand, the framework can lose the flexibility of the measurement procedures. We repeated some additions to the measurement parameters during the implementation because the request for such parameters varies with respective applications. For example, N-TAP currently doesn't provide the parameter of the ICMP packet size for measuring RTT, and the parameter may be important for some kinds of applications. This problem will be improved just by adding the parameter as an optional one. However, we must basically keep the framework simple but also extensible if needed on the development of N-TAP.

The deployment manner is an important factor as an infrastructure of network characteristics. At this time, we assume that N-TAP agents run on arbitrary end nodes because of the ease for obtaining end-to-end network characteristics, which is important for overlay networks. On the other hand, locating N-TAP nodes in respective administrative domains like DNS servers is possible, too. In the former case, we don't need any facility for operating N-TAP. However, there are some problems derived from the nature of a purely distributed system, e. g., the difficulty of maintaining the overlay network that the nodes frequently join to and remove from. In the latter case, we have to manage certain N-TAP nodes, and the status of N-TAP will be relatively easy to grasp. In that case, however, N-TAP needs some mechanisms of correction or estimation in order to obtain end-to-end characteristics because applications that utilize N-TAP are running on other nodes, not on N-TAP nodes. Summarizing such merits and demerits by implementing both models will constitute our future work as well.

With the current implementation, we've found that the information retrieval with *pull* style, which we chose because of its simplicity, often causes high load on both the application and the N-TAP agent. For example, an application that needs to check network characteristics continuously (e. g., for monitoring the change of the topology among overlay nodes) has to request to N-TAP frequently with making a TCP connection each time. In order to reduce such a burden, N-TAP should keep the TCP connection open rather than closing the connection after the response, or utilize UDP and define another lightweight protocol for the request. We also consider the information retrieval with *push* style. In *push* style, an application declares some conditions to N-TAP for receiving the notification, and then N-TAP gives network characteristics to the application only when these conditions are met. These alternatives will decrease the load derived from the repeated requests in exchange for losing the simplicity for handling network characteristics. We plan to support both styles of information retrieval in order to expand the purpose of use, while keeping the simplicity of the APIs.

4 Conclusion

In this paper, we posed the problem of the gap between measurement and overlay networks and explored a framework-oriented approach for obtaining network characteristics. We also discussed the challenges and future directions of N-TAP. The N-TAP project is still in the preliminary stages and many topics that should be studied and solved remain. However, we believe that our framework-oriented approach can provide the initiative toward a new trend in measurement platforms. It's time to show that measurement yields a profit to Internet users.

References

1. Dabek, F., Cox, R., Kaashoek, F., Morris, R.: Vivaldi: A Decentralized Network Coordinate System. In: Proc. of the 2004 ACM SIGCOMM Conference. (2004)
2. Donnet, B., Raoult, P., Friedman, T., Crovella, M.: Efficient Algorithms for Large-Scale Topology Discovery. In: Proc. of the 2005 ACM SIGMETRICS International Conference. (2005)
3. Masui, K., Kadobayashi, Y.: N-TAP: A Platform of Large-Scale Distributed Measurement for Overlay Network Applications. In: Proc. of the Second International Workshop on Dependable and Sustainable Peer-to-Peer Systems (DAS-P2P 2007). (2007)
4. Stoica, I., Morris, R., Liben-Nowell, D., Karger, D.R., Kaashoek, M.F., Dabek, F., Balakrishnan, H.: Chord: A Scalable Peer-to-Peer Lookup Protocol for Internet Applications. IEEE/ACM Transactions on Networking (TON) 11(1) (2003)
5. Madhyastha, H.V., Isdal, T., Piatek, M., Dixon, C., Anderson, T., Krishnamurthy, A., Venkataramani, A.: iPlane: An Information Plane for Distributed Services. In: Proc. of the 7th USENIX Symposium on Operating Systems Design and Implementation (OSDI '06). (2006)
6. Yalagandula, P., Sharma, P., Banerjee, S., Basu, S., Lee, S.J.: S^3: A Scalable Sensing Service for Monitoring Large Networked Systems. In: Proc. of the 2006 ACM SIGCOMM Workshop on Internet Network Management (INM '06). (2006)

Scanning Traffic at the Edge of a Cellular Network

Fabio Ricciato and Eduard Hasenleithner

Telecommunications Research Center Vienna (ftw.)
Donau-City Straße. 1, 1220 Vienna, Austria
{ricciato,hasenleithner}@ftw.at

Abstract. We report on the traffic and delay patterns observed at short time-scales at the edge of a cellular mobile network. We find that high-rate sequential scanners in the Internet are a common source of traffic impulses and introduce "noise" in the one-way delay measurements.

1 Introduction

The originary motivation for this work was to use one-way delay measurements as a basis to detect drifts and hidden problems between the measurement points: the emergence of delay values higher than the "physiological" level observed in the past can be taken as the symptom of capacity shortage or misfunctioning of some intermediate network element. We analyzed the one-way delays in the Core Network of an operational 3G mobile network in Austria, EU. We monitored all the GGSNs and the Edge Routers (ER) co-located at a single physical site. For a description of the network structure and the monitoring setting refer to [1]. We consider here only the downlink traffic, i.e. towards the Mobile Stations (MS). The one-way delay was extracted in post-processing with a similar methodology as in [2]: selected fields in the TCP/IP headers are hashed and matchings are seeked across different sections. Here we summarize our initial findings.

2 Findings

In Fig. 1 we plot the delay samples through the Gi section (from the peering links to the GGSNs, ref. [1, Fig. 1]) measured during 1 hour of moderate load [1]. As expected most of the samples take very low values, below 1 ms. However we observe also large delay values, up to 200 ms, concentrated into vertical lines (spikes) that are scattered uniformly across the whole day. The delay patterns internal to the GGSN looks similar (not reported here for space constraints, see [1, Fig. 3]). The simplest explanation is that delay spikes are associated to traffic bursts. In fact, the downlink traffic rate measured in bins of 1 sec yields

---

[1] Quantitative values are considered business-sensitive by the operator and subject to non-disclosure, e.g. absolute traffic volumes. For this reason the following graphs reporting absolute values have been rescaled by an arbitrary undisclosed factor.

S. Uhlig, K. Papagiannaki, and O. Bonaventure (Eds.): PAM 2007, LNCS 4427, pp. 269–272, 2007.

spikes (positive impulses) scattered across the whole day. By manual inspection we found that the primary cause for traffic spike are *sequential scanners*.

The probe packets originated by a sequential high-rate scanning source arrive to the local peering links in a pattern that is shaped by the address space allocation of the local network (see [1] for a detailed description). In Fig. 2 we report the packet rate measured at different network sections for the traffic originated by a single external IP address identified as a high-rate sequential scanner (call it "*S*"). The probe packets were arriving clustered into bursts of very short duration, corresponding to a relatively high bitrate (undisclosed). We found several active scanners, persistent and not, using different packet types (e.g. TCP SYN, UDP). When non-blocked ports are used the scanning bursts penetrate into the Core Network - note that public addressing is used for the MS.

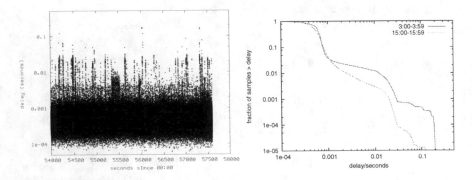

Fig. 1. One-way downlink delays on Gi, from peering links to GGSNs

Comparing Figg. 2(a) and 2(b) we see that a sensible fraction of the probe packets (up to 80%) is lost in the Gi path (ER → GGSN) due to micro-congestion caused by the the scanning traffic itself (discussed below). Comparing 2(b) and 2(c) we see that in each timebin only a fraction of the packets seen on Gi reach the Gn links, i.e. passes the GGSN. In fact the latter forwards only the packets directed to active IP addresses, i.e. currently assigned to MS within active connections (PDP-contexts), hence the penetration of scanning traffic into Gn is subject to time-of-day effect: low in the night, maximum at the peak hour when

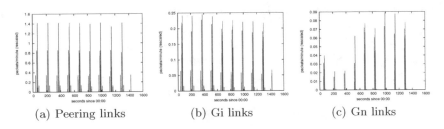

(a) Peering links   (b) Gi links   (c) Gn links

Fig. 2. Daily packet rate (rescaled) for one scanning source, 1 min bins

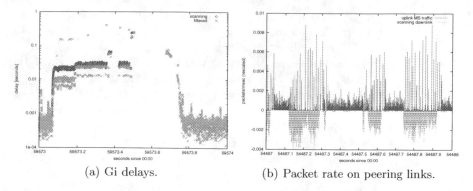

(a) Gi delays.   (b) Packet rate on peering links.

Fig. 3. Microscopic patterns during scanning bursts

more MS are active. Notably a fraction of MS respond to incoming probes (e.g. with ICMP "port unreachable") thus causing backscatter traffic in uplink.

Micro-congestion. Scanning bursts were clipped inside the Gi section because hitting the capacity limit of some internal link. To see that we zoom into a sample scanning burst from source S. We divide the delay samples into two groups: scanning packets (identified by the IP address of S) and other traffic ("filtered"). We plot them separately in Fig. 3(a) against the time-axis. Let us focus on the "scanning" samples: the delay pattern is consistent with the presence of a buffer that fills up rapidly (initial slope) and then remains persistently saturated (plateau). After the fill-up phase most arriving packets are lost, while those entering the buffer experience an approximately constant delay equal to the buffer depletion time (≈ 20 ms in this case). After this phase we observe a *cluster* of delay samples considerably larger, around 200 ms, followed by an empty period of ≈ 200 ms wherein no packets are seen at all. This pattern is consistent with the so-called "coffee-break" event [2], i.e. a temporary interruption of the packet forwarding process at some intermediate router. In our traces such events only occurr during large scanning bursts, suggesting that the observed coffee-breaks are *not* to due to "normal" router dynamics as in [2] but rather a symptom of short-term CPU congestion. Note also that in Fig. 3(a) the whole delay pattern described for the scanning packets is followed by the other traffic as well ("filtered" series). This indicates that the micro-congested resources (buffer, CPU) are shared by all traffic. In other words, the scanning traffic is causing a small impairment to the other legitimate traffic.

Besides micro-congestion we observed a number of other interesting phenomena somehow related to the presence of scanning traffic.

Traffic Notches. The total traffic rate observed in the core network (Gi, Gn links) displays some *notches*, i.e. negative impulses, when measured at small timescales (1 sec bins, see [1, Fig. 4]). Further investigations into a few sample cases revealed a close correlation with the scanning process, suggesting that high-rate scanning bursts can locally reduce the arrival rate of legitimate traffic.

Uplink packing. The microscopic analysis revealed that the presence of scanning traffic in downlink has an impact on the traffic pattern in *uplink* as well. In Fig. 3(b) we draw the uplink traffic originated by all the MS as observed at the peering links, measured as packet counts in timebins of 1 ms. On the negative axis we draw the count of scanning packets seen in downlink. Note that during the scanning periods the uplink packets are transmitted into discontinuous bursts, corresponding to large distinct spikes in the packet count. A possible explanation is that the CPU at some internal node is kept busy by the scanning traffic for a few milliseconds, thus starving the uplink forwarding process. During such vacancy periods the arriving uplink packets are buffered, then when the CPU returns available for the uplink stream they are forwarded at once, "packed" into a single burst. Notably most of uplink packets are TCP ACK, which might in principle lead to some sort of TCP synchronization effect at small time-scale.

3   Conclusions and Future Work

We found that Internet traffic contains large bursts of packets originated by high-rate sequential scanning activities. Large access network employing public IP addressing, including 3G cellular networks, are permeable to scanning bursts. The impact of such activities - and of unwanted traffic in general [3] - onto the underlying network infrastructure might be not yet well understood. The initial observations reported here, still specific to the particular network under study, suggest that the network dynamics triggered by such traffic are non-trivial and worth further investigations. We showed that scanning traffic causes a sort of "shot-noise" into the delay measurements due to micro-congestion events. Such noise complicates the task of using the delay statistics to validate the health status of the network elements, and in general should be taken into account when interpreting delay measurements from the real-world.

References

1. F. Ricciato, E. Hasenleithner. Observations at short time-scales from the edge of a cellular data network. *Technical Report FTW-TR-2007-001*, January 2007. Available online from [4].
2. K. Papagiannaki, S. Moon, C. Fraleigh, P. Thiran, C. Diot. Measurement and Analysis of Single-Hop Delay on an IP Backbone Network. *IEEE JSAC*, 21(6), August 2003.
3. F. Ricciato. Unwanted Traffic in 3G Networks. *ACM Computer Communication Review*, 36(2), April 2006.
4. DARWIN home page http://userver.ftw.at/~ricciato/darwin.

Author Index

Lecture Notes in Computer Science

For information about Vols. 1–4329

please contact your bookseller or Springer

Vol. 4377: M. Abe (Ed.), Topics in Cryptology – CT-RSA 2007. XI, 403 pages. 2006.

Vol. 4376: E. Frachtenberg, U. Schwiegelshohn (Eds.), Job Scheduling Strategies for Parallel Processing. VII, 257 pages. 2007.

Vol. 4374: J.F. Peters, A. Skowron, I. Düntsch, J. Grzymała-Busse, E. Orłowska, L. Polkowski (Eds.), Transactions on Rough Sets VI, Part I. XII, 499 pages. 2007.

Vol. 4373: K. Langendoen, T. Voigt (Eds.), Wireless Sensor Networks. XIII, 358 pages. 2007.

Vol. 4372: M. Kaufmann, D. Wagner (Eds.), Graph Drawing. XIV, 454 pages. 2007.

Vol. 4371: K. Inoue, K. Satoh, F. Toni (Eds.), Computational Logic in Multi-Agent Systems. X, 315 pages. 2007. (Sublibrary LNAI).

Vol. 4370: P.P Lévy, B. Le Grand, F. Poulet, M. Soto, L. Darago, L. Toubiana, J.-F. Vibert (Eds.), Pixelization Paradigm. XV, 279 pages. 2007.

Vol. 4369: M. Umeda, A. Wolf, O. Bartenstein, U. Geske, D. Seipel, O. Takata (Eds.), Declarative Programming for Knowledge Management. X, 229 pages. 2006. (Sublibrary LNAI).

Vol. 4368: T. Erlebach, C. Kaklamanis (Eds.), Approximation and Online Algorithms. X, 345 pages. 2007.

Vol. 4367: K. De Bosschere, D. Kaeli, P. Stenström, D. Whalley, T. Ungerer (Eds.), High Performance Embedded Architectures and Compilers. XI, 307 pages. 2007.

Vol. 4366: K. Tuyls, R. Westra, Y. Saeys, A. Nowé (Eds.), Knowledge Discovery and Emergent Complexity in Bioinformatics. IX, 183 pages. 2007. (Sublibrary LNBI).

Vol. 4364: T. Kühne (Ed.), Models in Software Engineering. XI, 332 pages. 2007.

Vol. 4362: J. van Leeuwen, G.F. Italiano, W. van der Hoek, C. Meinel, H. Sack, F. Plášil (Eds.), SOFSEM 2007: Theory and Practice of Computer Science. XXI, 937 pages. 2007.

Vol. 4361: H.J. Hoogeboom, G. Păun, G. Rozenberg, A. Salomaa (Eds.), Membrane Computing. IX, 555 pages. 2006.

Vol. 4360: W. Dubitzky, A. Schuster, P.M.A. Sloot, M. Schroeder, M. Romberg (Eds.), Distributed, High-Performance and Grid Computing in Computational Biology. X, 192 pages. 2007. (Sublibrary LNBI).

Vol. 4358: R. Vidal, A. Heyden, Y. Ma (Eds.), Dynamical Vision. IX, 329 pages. 2007.

Vol. 4357: L. Buttyán, V. Gligor, D. Westhoff (Eds.), Security and Privacy in Ad-Hoc and Sensor Networks. X, 193 pages. 2006.

Vol. 4355: J. Julliand, O. Kouchnarenko (Eds.), B 2007: Formal Specification and Development in B. XIII, 293 pages. 2006.

Vol. 4354: M. Hanus (Ed.), Practical Aspects of Declarative Languages. X, 335 pages. 2006.

Vol. 4353: T. Schwentick, D. Suciu (Eds.), Database Theory – ICDT 2007. XI, 419 pages. 2006.

Vol. 4352: T.-J. Cham, J. Cai, C. Dorai, D. Rajan, T.-S. Chua, L.-T. Chia (Eds.), Advances in Multimedia Modeling, Part II. XVIII, 743 pages. 2006.

Vol. 4351: T.-J. Cham, J. Cai, C. Dorai, D. Rajan, T.-S. Chua, L.-T. Chia (Eds.), Advances in Multimedia Modeling, Part I. XIX, 797 pages. 2006.

Vol. 4349: B. Cook, A. Podelski (Eds.), Verification, Model Checking, and Abstract Interpretation. XI, 395 pages. 2007.

Vol. 4348: S.T. Taft, R.A. Duff, R.L. Brukardt, E. Ploedereder, P. Leroy (Eds.), Ada 2005 Reference Manual. XXII, 765 pages. 2006.

Vol. 4347: J. Lopez (Ed.), Critical Information Infrastructures Security. X, 286 pages. 2006.

Vol. 4346: L. Brim, B. Haverkort, M. Leucker, J. van de Pol (Eds.), Formal Methods: Applications and Technology. X, 363 pages. 2007.

Vol. 4345: N. Maglaveras, I. Chouvarda, V. Koutkias, R. Brause (Eds.), Biological and Medical Data Analysis. XIII, 496 pages. 2006. (Sublibrary LNBI).

Vol. 4344: V. Gruhn, F. Oquendo (Eds.), Software Architecture. X, 245 pages. 2006.

Vol. 4342: H. de Swart, E. Orłowska, G. Schmidt, M. Roubens (Eds.), Theory and Applications of Relational Structures as Knowledge Instruments II. X, 373 pages. 2006. (Sublibrary LNAI).

Vol. 4341: P.Q. Nguyen (Ed.), Progress in Cryptology - VIETCRYPT 2006. XI, 385 pages. 2006.

Vol. 4340: R. Prodan, T. Fahringer, Grid Computing. XXIII, 317 pages. 2007.

Vol. 4339: E. Ayguadé, G. Baumgartner, J. Ramanujam, P. Sadayappan (Eds.), Languages and Compilers for Parallel Computing. XI, 476 pages. 2006.

Vol. 4338: P. Kalra, S. Peleg (Eds.), Computer Vision, Graphics and Image Processing. XV, 965 pages. 2006.

Vol. 4337: S. Arun-Kumar, N. Garg (Eds.), FSTTCS 2006: Foundations of Software Technology and Theoretical Computer Science. XIII, 430 pages. 2006.

Vol. 4336: V.R. Basili, D. Rombach, K. Schneider, B. Kitchenham, D. Pfahl, R.W. Selby, Empirical Software Engineering Issues. XVII, 193 pages. 2007.

Vol. 4335: S.A. Brueckner, S. Hassas, M. Jelasity, D. Yamins (Eds.), Engineering Self-Organising Systems. XII, 212 pages. 2007. (Sublibrary LNAI).

Vol. 4334: B. Beckert, R. Hähnle, P.H. Schmitt (Eds.), Verification of Object-Oriented Software. XXIX, 658 pages. 2007. (Sublibrary LNAI).

Vol. 4333: U. Reimer, D. Karagiannis (Eds.), Practical Aspects of Knowledge Management. XII, 338 pages. 2006. (Sublibrary LNAI).

Vol. 4332: A. Bagchi, V. Atluri (Eds.), Information Systems Security. XV, 382 pages. 2006.

Vol. 4331: G. Min, B. Di Martino, L.T. Yang, M. Guo, G. Ruenger (Eds.), Frontiers of High Performance Computing and Networking – ISPA 2006 Workshops. XXXVII, 1141 pages. 2006.

Vol. 4330: M. Guo, L.T. Yang, B. Di Martino, H.P. Zima, J. Dongarra, F. Tang (Eds.), Parallel and Distributed Processing and Applications. XVIII, 953 pages. 2006.